2ND Edition

Hydraulics

By

Dr M. Zaher

Auckland, New Zealand

July 2014

PREFACE

Hydraulics is an applied division of fluid mechanics covering a specific range of engineering problems and solutions. It studies the laws of equilibrium and motion of fluids, also their application to practical problems.

The principle concern of hydraulics is the fluid flow constrained by surrounding surfaces, i.e. flow in open and closed channels/conduits, including rivers and canals, as well as pipes, nozzles and hydraulic machine elements.

Hydraulic is mainly concerned with the internal flow of fluids, investigating what might be called 'internal' problems, in distinction to 'external' problems involving flow of continuous medium about submerged bodies, as is the case of a solid body moving in water or in air. Such external problems are treated within discipline like hydrodynamics and aerodynamics in connection with aircraft and ship design.

The difference between a liquid and a gas is that the power tends to contain itself to a specific shape in small quantities and makes a free surface in large volumes. An important property of liquids is that pressure or temperature changes have practically no effect on their volume i.e. for all practical purposes they are regarded as incompressible. Gases, on the other hand, contract under pressure and expand infinitely in the absence of pressure i.e. they are highly compressible. Despite this difference however, with certain conditions the law of motion of liquids and gases are practically identical. One such condition is low velocity of the gas flow as compared with the speed of sound through gas.

The science of hydraulics concerns itself mainly with the motion of liquids. The internal flow of gases is studied only insofar as the velocity of flow is much less than that of sound; consequently, their compressibility can be disregarded. Such cases are frequently encountered in engineering, as for example, in the flow of air in ventilation systems and in gas mains.

Investigation of the flow of liquids is more complicated than studying the motion of rigid bodies. In the mechanics of rigid bodies one deal with systems of rigidly connected particles, in fluid mechanics the objective is to investigate a medium consisting of a multitude of particles in constant relative motion.

With Galileo's principle concept, it then became easier to study the motion of remote bodies than of a stream running nearby. Because of these difficulties, fluid mechanics as a science developed along two different paths:

The first was the purely theoretical path of precise mathematical analysis based on the laws of mechanics. This led to the formation of theoretical hydromechanics, which for a long time existed as an independent discipline. Its methods provided an attractive and effective means of scientific research. A theoretical analysis of fluid motion, however, encountered many obstacles, which didn't always answer questions arising in real situations.

The urgent requirements of engineering practice soon gave rise to a new science of fluid motion. This was what we know today as Hydraulics, in which researchers adopted the path of extensive experimentation and the accumulation of factual data for application to engineering problems. At the start, hydraulics was a purely empirical science. Today - whenever necessary - it employs the methods of theoretical hydromechanics for the solution of various problems and experiments are widely used to verify the validity of the problem.

Gradually the difference in the methods employed by both disciplines has faded away and reference is made to Hydraulics.

Today, the method of investigating fluid flow in hydraulics is essentially a flow. The phenomenon under investigation is first simplified and idealised, while the laws of theoretical mechanics are applied. The results are then compared with experimental data. The discrepancies are then established and the theoretical formulas and solutions are adjusted to make them suitable for practical application.

Various practical phenomenon are involved in defying their theoretical analysis and those investigated in hydraulics on the sole basis of experimental measurement, the results being expressed as an empirical formulas. Thus it is valid to call Hydraulics a semi-empirical science.

At the same time Hydraulics is considered an applied engineering science in so far as it functions in accordance with the demands of life and is widely used in engineering. Hydraulics provides the methods of calculation and design of a wide range of hydraulic structures (dams, canals, weirs and pipelines), machinery (pumps, turbines and hydraulic machines) and other devices used in many branches of engineering.

Hydrology: Chapters (15) to (17) of this book is devoted to the introduction of engineering hydrology, which is defined as the science that treats waters on earth, their occurrence, circulation, distribution, their chemical and physical properties and their environment including their relation to living things. In short: what happens to the rain is the basis of the definition of the science of hydrology. Hydrology also got a variety of practical applications which should not be treated as a pure science.

During the development of this book the most frequently asked questions were: "What type of audience will the material address?" The arbitrary division is placed between the service and maintenance technicians and engineering personnel.

This book introduces the fundamentals of Hydraulics and engineering Hydrology. It is intended as a text for undergraduate students in engineering. The book contains material suitable for a particular course, which can safely be entrusted to the instructor. It is not a predigest of some venerable work, but a textbook whose merits and deficiencies are peculiarly its own. Reading the book itself can best assess it.

Dr M. Zaher
Auckland - New Zealand

TABLE OF CONTENTS

1. Fluid Properties

1.1 Introduction

The conditions that govern movement of bodies along the x-y-z plans are properties such as friction, viscosity, and pressure.

Mechanics is a science concerned with the motion of bodies and the conditions governing such motion. The subject of mechanics is frequently divided into two general sections known as 'Kinematics' and 'Dynamics'.

Hydraulics is the branch of engineering-science which deals with liquids (i.e. water, oil, ..), at rest or in motion. The hydraulic machines are the branch of engineering-science which deals with the machines run by water, under some head or raising the water to higher levels.

1.2 Liquids and their Properties

There is no difficulty in distinguishing a liquid from a solid or gas. A solid has a definite shape, which it retains, until some external force is applied to alter its shape. Liquid takes the shape of the vessel into which it is poured. Gas completely fills up the vessel which contains it.

Among the liquids, water is the one mostly dealt with. It has the following properties:

a) Density,
b) Compressibility,
c) Viscosity, and
d) Surface tension.

1.3 Density

The density of a liquid is the mass per unit volume at a standard temperature and pressure. Mass density denoted by ρ (rho).

1.4 Specific Weight of Water

Is defined as the weight per unit volume at standard temperature and pressure. Weight density or specific weight is denoted by (w):

$$w = \rho \ g \ kg/m^3 \ (g = \text{gravitational acceleration - } m/s^2)$$

1.5 Specific Gravity of Water

Is defined as the ratio of specific weight, to that of a standard substance at a standard temperature (usually taken as pure water at 4°C).

Sp.gr. = specific weight of liquid / Specific weight of pure water = w_{liquid} / w_{water}

1.6 Compressibility of Water

Defined as the variation in its volume, with the variation of pressure. This variation in liquids is very small and is neglected in calculations. Water is considered as an incompressible liquid.

1.7 Surface Tension of Water

Is the property of the liquid to resist tensile stress. It is a result of cohesion between the molecules at the surface of a liquid.

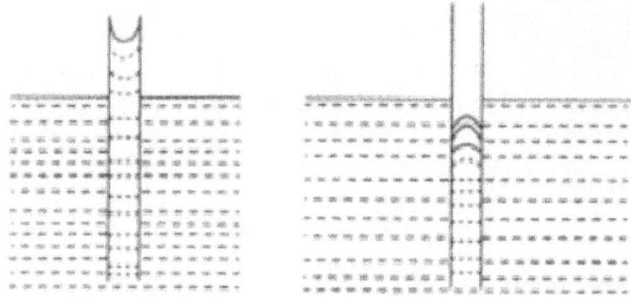

Fig. (1-1) Water upward concave, mercury depressed with upward convex

The effect of surface tension is usually neglected in hydraulics calculations.

1.8 Capillary of Water

Why mercury does not wet glass? Because the cohesion between the mercury molecules is much greater than the adhesion between the tube and mercury molecules.

The phenomenon of rising water in the tube of smaller diameter is called the capillary rise:

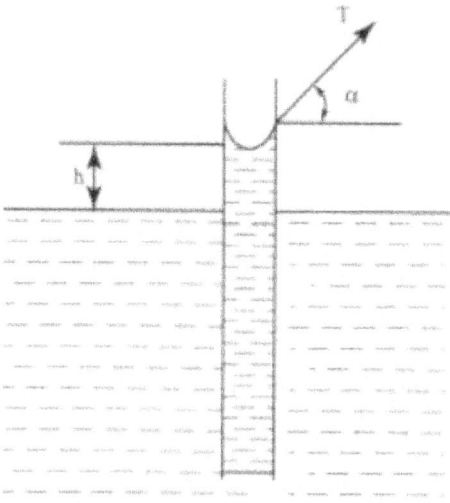

Fig. (1-2) Capillary rise

Let, h = height of capillary rise [mm],
   d = diameter of the capillary tube [mm],
   α = angle of contact of the water surface [degrees/rad], and
   T = the force of surface tension per unit length of the periphery of the capillary tube
   [kg/mm].

Weight of water column in the tube above the water surface acting downwards
$$= w\ h\ \pi/4\ (d)^2 \qquad\qquad (1\text{-}1)$$

Vertical component of the force of surface tension
$$= T\ \pi\ d\ \sin\alpha \qquad\qquad (1\text{-}2)$$

Since the downward weight of the water column is balanced by the vertical component of the force of surface tension, therefore it is possible to equate (1-1) and (1-2):
$$w\ h\ \pi/4\ (d)^2 = T\ \pi\ d\ \sin\alpha$$
Hence,
$$h = 4\ T\ \sin\alpha\ /\ w\ d$$

1.9 Viscosity of Liquids

We see that flow of liquids, such as alcohol or water, is much easier than thick liquids such as syrup or heavy oil. It is thus obvious that each liquid possesses some property, which controls its rate of flow. This property is termed as 'viscosity' and is essential property of liquid.

1.10 Mass and Weight

(i) Mass: is the amount of matter contained in a given body, and does not vary with the change in its position on earth's surface [kg]. Use lever balance.

(ii) Weight: It is the amount of pull, which the earth exerts upon a given body. Weights vary with position on earth surface (elevations). Weight is a force [kg-force]. Use spring balance.

1.11 Laws of Motion (dynamics)

(i) Newton's first law of motion: 'All bodies continue in their state of rest or uniform motion, in a straight line, unless acted upon by some external force.'

(ii) Newton's second law of motion: 'The rate (time) of change of momentum (= mass x velocity) is directly proportional to the acting force and takes place in the same direction in which this force acts.'

Let, m = mass of the body [kg],
   P = force or rate of change of momentum [kg],
   u = initial velocity of the body [m/s],
   v = final velocity of the body [m/s], and
   t = time taken, by the body, in changing its velocity from u to v [s].

According to Newton's second law of motion:
$$P \alpha (mv - mu) / t \alpha m(v - u) / t \alpha m f \text{ (where, } f = \text{acceleration} = v\text{-}u/t) = k m f$$

where, k is a constant. To simplify, the unit of force adopted produces a unit acceleration to a body of unit mass:
$$P = m f = \text{mass x acceleration}$$
(iii) Newton's third law of motion: 'To every action, there is always an equal and opposite reaction.' - motion of a rocket.

$$\text{Work = Force x distance = P x s}$$
$$\text{Power = Work/Time HP}$$
$$\text{Energy = is the capacity to do work [kg-m]}$$

1.12 Laws of Conservation of Energy

Energy can neither be created nor destroyed, though it can be transformed from

1.13 SI Units / Engineering Units

There are six basic SI units, and the units of other thermodynamic quantities are derived from these basic units. The basic SI units are as follows:

Length: The meter (m) is defined as the length of the path travelled by light in vacuum during a time interval of 1/299792458 of a second.

Mass: The kilogram (kg) mass is equal to the mass of the international prototype of the kilogram. This international prototype is made of platinum-iridium metal and has been kept at the International Bureau of Weights and Measures, Severs, France.

Time: Time seconds (s) is the duration of 9,192,631,770 periods of radiation corresponding to the transition between the two hyper line levels of the ground state of the caesium 133 atoms.

Electric current: The electric current is that quantity of current which, if maintained in two straight parallel conductors placed 1 meter apart in vacuum, would produce between these conductors a force equal to 2×10^{-7} newton per meter of length.

Temperature: The temperature in kelvin (K) is 1/273.6 times the thermodynamic temperature of the triple point of water.

Force: The unit of force is newton (N). One newton force is that force which, when applied to a body having a mass of one kilogram, gives it an acceleration of one meter per second per second (I m/s^2). Mathematically force can be expressed as: F = (C) (m) (a)
Where, F = force in newton (N), m = mass in kilograms (kg), a = acceleration in meters per second per second (m/s^2) and C = proportionality constant. The SI units of force, newton, is derived assuming this constant as unity, i.e.
$$1 \text{ N} = (1 \text{ kg}) \times (1 \text{ m/s}^2) = 1 \text{ (kg. m)}/s^2$$
The cgs unit of force is dyne, which is the force exerted on 1 gram mass for 1 cm/s^2 acceleration, i.e.
$$1 \text{ dyne} = 1 \text{ g.cm/s}^2$$
Or,
$$1 \text{ N} = 10^5 \text{ g.} \frac{cm}{s^2} = 10^5 \text{ dyne}$$

Fluid pressure: It is the rate of change of momentum of the fluid particles per unit time per unit area.

Atmospheric pressure: The earth is surrounded by an envelope of atmosphere or air that extends upwards from the earth's surface to height of 80 km or more. Since air has mass, subject to the action of gravity, it exerts pressure known as 'atmospheric pressure'.
Suppose there is an air column with its cross-section 1 m^2, starting from the surface of the earth at sea level and extending to the upper limit of atmosphere. Such an air column is supposed to have exerted on 1 m^2, therefore the pressure exerted by the atmosphere at sea level is 101.325 N/m^2. This atmospheric pressure id considered as standard and is used in all the calculations.
The atmospheric pressure changes somewhat with temperature, humidity and altitude. It can be seen that one standard atmosphere is given by:
    1 atm = 1.01325 bar = 1.033 kg f/cm^2 = 760 mm Hg = 10.336 meter of water column
The pressure of any fluid or system measured with the help of instruments is known as 'gauge pressure'. But thermodynamic investigations are connected with absolute pressure.

Energy and work: Energy is an idea, central to the development of all branches of science and engineering. A fundamental postulate of thermodynamics is that matter possesses energy (which can be in several forms) and energy is conserved.

Energy is nothing but the ability to do work. It is always said that energy exists in transition only. The unit of work or energy is obtained from the product of force and distance. The SI unit of work or energy is newton-meter (N-m) or joule (J).

$$1 \text{ N-m} = 1 \text{ J} = 10^7 \text{ erg} = 10^7 \text{ dyne} - \text{cm}$$

The conversion of MKS unit of energy to SI unit is:

$$1 \text{ kcal} = 4186.8 \text{ N-m} = 4.1868 \text{ kJ} = 3.968 \text{ Btu}$$

Or, $$1 \text{ kJ} = 0.239 \text{ kcal} = 0.948 \text{ Btu}$$

Power: Power is the rate of doing work. The unit of power is watt (W). One watt power is defined as doing work at the rate of one joule per second (J/s). In electrical engineering, one watt power is defined as:

$$1 \text{ W} = 1 \text{ (volt)} \times 1 \text{ (ampere)} = 1 \text{ J/s}$$

The relationship between horsepower and watt is as follows:

$$1 \text{ hp (empirical)} = 746 \text{ N-m/s or J/s or W}$$

$$1 \text{ hp (metric)} = 75 \text{ kg/m/s} = 75 \times 9.80665 \text{ N-m/s} = 736 \text{ N-m/s or W}$$

Further, the units of energy can be obtained from that of power. Thus,

$$1 \text{ J} = 1 \text{ W-s}$$

$$1 \text{ kWh} = 1000 \times 3600 \text{ J} = 3600 \text{ kJ}$$

Enthalpy: Enthalpy is nothing but the sum of internal energy and the product of pressure and volume. It is denoted by H and measured in joules. Mathematically:

$$H = U + pV$$

where,

U = internal energy (J)

p = pressure (N/m^2)

V = volume (m^3)

The specific enthalpy is denoted by 'h' and measured in J/kg.

Entropy: Entropy is difficult to define and can be said to be the property of the system such that for any reversible process between the state points 1 and 2, its change is given by:

$$\delta s = \int_1^2 (\delta Q/T)_{rev}$$

The SI unit of specific entropy is kJ/kg-K. It can also be expressed as 1 kcal/kg°C = 4.186 kJ/kg-K

Check Your Knowledge

1. Distinguish between a solid and a liquid?
2. Differentiate between density, specific weight and specific gravity of a liquid.
3. Show that rise of liquid in a capillary tube given by the equation:

$$h = (2 \, \sigma \cos\Theta) / (w \, r)$$

where, h = rise of liquid above normal level,

σ = surface tension,

Θ = angle, which the capillary rise surface makes with the vertical,

w = specific weight of liquid, and

r = radius of capillary tube.

4. What are the various systems of fundamental units? Describe the international SI units
5. What is the difference between Mass and Weight? Which is measured by a beam balance and which by a spring balance?
6. State Newton's three laws of motion.

2. Hydrostatics

2.1 Fundamentals

Hydrostatic refers to a study of the conditions under which a particle or body remains at rest. They are moving sufficiently slowly with no relative motion between adjacent parts of the body.

If a particle or body is at rest, under a given set of forces, the forces are said to be in equilibrium with no shear stresses considered, while there are only pressure forces that act perpendicular to any surface. The body is said to be in 'static equilibrium'. The absence of shear means that friction need not be considered.

2.2 Hydrostatic Pressure and Force

When fluid is contained in a vessel, it exerts force at all points on the sides and bottom of the container. This force per unit area is called 'pressure'.

If 'F' is the force acting on an area 'A', then intensity of pressure (generally termed pressure) is:

$$p = F / A \qquad\qquad (2\text{-}1)$$

The direction of this pressure is always at right angles to the surface with which the fluid at rest comes in contact.

2.2.1 Pressure head: Consider a vessel containing some liquid. The liquid will exert pressure on all sides and on the bottom of the vessel:

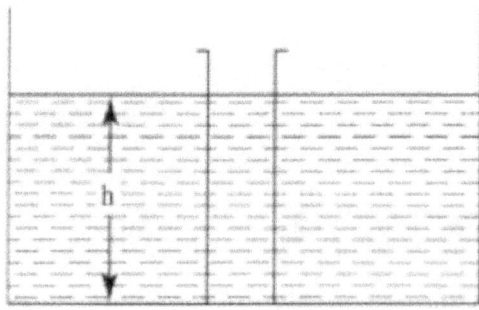

Fig (2-1) Cylinder in a vessel

Insert a cylinder in up-right position. The weight of liquid contained in the cylinder = w h A,

where, w = specific weight of the liquid,

      h = height of liquid in cylinder, and

      A = area of cylinder base.

The pressure, at the bottom of the cylinder will be due to the weight of the liquid contained in the cylinder.

Assume the pressure to be 'p':
Then, p = weight of liquid in the cylinder/Area of the cylinder base = w h A / A = w h

This equation shows that the intensity of pressure at any point in a liquid is proportional to its depth from the surface (as 'w' is constant for a given liquid).

The pressure can be expressed in either of the following ways:
(i) As a force per unit area (i.e. kg/m^2)
(ii) As a height of equivalent liquid column.

Pascal's law: States that the intensity of pressure at any point in a fluid at rest is the same in all directions, as shown in Figure (2-2):

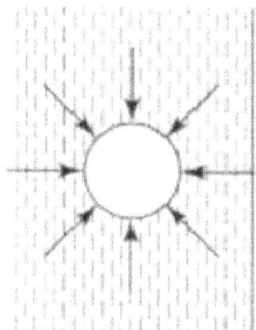

Fig (2-2)   Pressure at a point

It also states that: 'The pressure in a fluid at rest is independent of direction as long as there are no shearing stresses present'.

This law can be generalised to a moving fluid, but if shearing stress are present, it is no longer true.

Consider a wedge of water as shown in Figure (2-3) below:

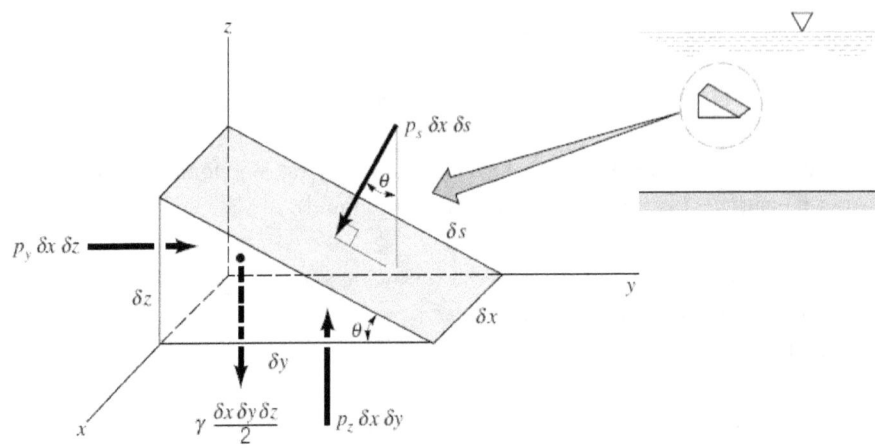

Fig (2-3) Pressures on wedge of water

The forces acting on the wedge are in equilibrium, hence:

$$\Sigma \, F_y = p_y \, \delta_x \, \delta_z - p_s \, \delta_x \, \delta_s \sin\Theta = 0$$

$$\Sigma \, F_z = p_z \, \delta_x \, \delta_y - p_s \, \delta_x \, \delta_s \cos\Theta - \gamma \, (\delta_x \delta_y \delta_z / 2) \; = 0$$

Since, volume of wedge $= \delta_x \delta_y \delta_z / 2$, and from figure (2-3) geometry we get:

$$\delta_y = \delta_s \cos\Theta; \qquad \delta_z = \delta_s \sin\Theta,$$

Hence,
$$p_y \, \delta_x \, \delta_z = p_s \, \delta_x \, \delta_z$$
$$p_z \, \delta_x \, \delta_y - p_s \, \delta_x \, \delta_y = \gamma \, (\delta_x \delta_y \delta_z / 2)$$

Therefore,
$$p_y = p_s$$

Similarly,
$$p_z \, \delta_x \, \delta_y - p_s \, \delta_x \, \delta_y = \gamma \, (\delta_x \delta_y \delta_z / 2)$$

Therefore,
$$p_z - p_s = \gamma \, \delta_z / 2$$

Where,

p_y Is the pressure of the left side, p_y and the inclined plane p_s are equal.

p_z Is the pressure of the bottom side, p_z and the inclined plane p_s are equal when the size of the wedge goes to zero.

2.2.2 Atmospheric pressure: It has been established scientifically that air possesses weight. Subsequently, it was also thought that the air, due to weight, must exert some pressure on the surface of the earth. Since air is compressible; therefore its density is different at different heights. The density of air has also been found to vary from time to time due to the changes in its temperature and humidity. It is thus obvious that due to the difficulties the atmospheric pressure (which is due to weight of the atmosphere or air above the surface of the earth) cannot be calculated as is done in the case of liquids. However, it is measured by the height of the column of liquid that it can support.

It has been observed that at sea level the pressure exerted by the column of air of one square cm cross sectional area and of height equal to that of the atmosphere is 1.03 kg. Thus we may say that the atmospheric pressure at the sea level is 1.03 kg/cm^2. It can also be expressed as 10.3 meters of water, in terms of equivalent water column or 76 cm of mercury in terms of equivalent mercury column.

2.2.3 Gauge pressure: It is the pressure, measured with the help of a pressure measuring instrument, in which the atmospheric pressure is taken as datum; or in other words the atmospheric pressure on the gauge seal is marked as zero. Generally, this pressure is above the atmospheric pressure.

2.2.4 Absolute pressure: It is pressure equal to the algebraic sum of atmospheric and gauge pressure. A little consideration will show that if the gauge pressure is minus (as the case of vacuum or suctions); the absolute pressure will be atmospheric pressure minus gauge pressure.

2.3 Total Pressure on Immersed Surface

The total pressure, on immersed surface, may be defined as the total pressure exerted by the liquid on it.

Mathematically,
$$P = p_1 a_1 + p_2 a_2 + p_3 a_3 + \ldots$$
Where,
P   = total force,
$P_1 P_2 P_3 \ldots$ = intensity of pressure on different strips of the surface, and
$a_1 a_2 a_3 \ldots$ = Areas of corresponding strips.

When we dive in a swimming pool, we feel some uneasiness. As we dive deeper and deeper, we feel more and more uneasiness. This uneasiness is due to the weight of water above us.

Now we shall discuss the total pressure exerted by a liquid on an immersed surface. The position of an immersed surface may be:
(1) horizontal, (2) vertical, or (3) inclined.

2.3.1 Total pressure on a horizontal immersed surface: Consider a plane horizontal surface immersed in a liquid as shown in Figure (2-4):

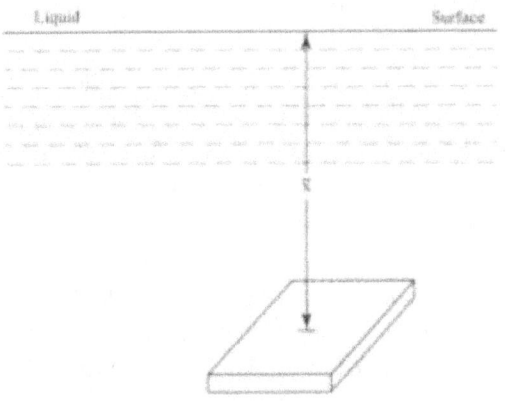

Fig (2-4) Horizontal surface immersed in liquid

Let,
w = sp. weight of the liquid,
A = area of the immersed surface, and
$\bar{x}$ = depth of the horizontal surface from the liquid level.

Total pressure on the surface:

P = weight of the liquid above the immersed surface = sp. weight of liquid x volume of liquid
   = sp. weight of liquid x area of surface x depth of liquid = w A $\bar{x}$

2.3.2 Total pressure on a vertical immersed surface: Consider a plane vertical surface immersed in a liquid as shown in Figure (2-5):

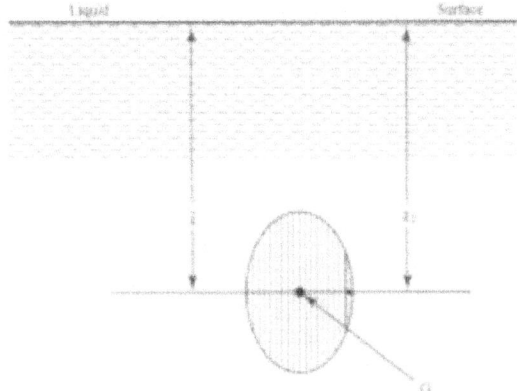

Fig (2-5) Plane vertical surface immersed in liquid

Divide the whole surface into a number of small parallel strips:

Sp. weight of liquid = w

a_1, a_2, a_3 ....   = areas of the strips,

x_1, x_2, x_3 ....   = depth of corresponding strips from the liquid surface,

Pressure on the first strip = $w\, a_1\, x_1$

Similarly,

Pressure on the second strip = $w\, a_2\, x_2$, and

Pressure on the third strip   = $w\, a_3\, x_3$

Total pressure on the surface P = $w\, a_1\, x_1 + w\, a_2\, x_2 + w\, a_3\, x_3 = w\,(a_1\, x_1 + a_2\, x_2 + a_3\, x_3) =$
$$= w\, A\, \bar{x}$$

Where,

A = area of the surface, and

$\bar{x}$ = depth of centre of gravity of the surface from the liquid surface.

2.3.3 Total pressure on an inclined immersed surface: Consider a plain inclined surface, immersed in liquid as shown in Figure (2-6):

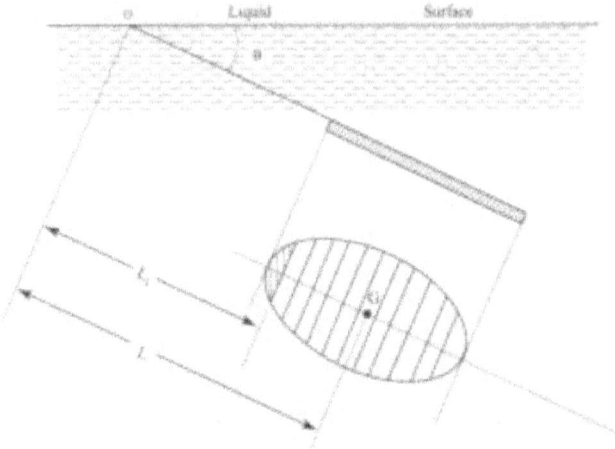

Fig (2-6) Inclined surface immersed in liquid

Divide the whole surface into a number of small parallel strips:

Let,

w　　　　　　　　= specific weight of liquid,

a_1, a_2, a_3 .... = areas of the strips,

L_1, L_2, L_3 .... = distance of corresponding strips from 0,

Θ　　　　　　　= angle at which the immersed surface is inclined with the liquid surface.

Pressure on the first strip = $w\, a_1\, L_1\, \sin\Theta$

Similarly,

Pressure on the second strip = $w\, a_2\, L_2\, \sin\Theta$, and

Pressure on the third strip　= $w\, a_3\, L_3\, \sin\Theta$

The total pressure on the surface:

$P = w\, a_1\, L_1\, \sin\Theta + w\, a_2\, L_2\, \sin\Theta + w\, a_3\, L_3\, \sin\Theta + \ldots$

　= $w \sin\Theta\, (a_1\, L_1 + a_2\, L_2 + a_3\, L_3 + \ldots.)$

　= $w\, A\, L\, \sin\Theta$

Where,

L = distance of the centre of gravity of the surface from 0,

A = area of the surface,

$\bar{x}$ = depth of centre of gravity of the surface from the liquid surface,

$P = w\, A\, \bar{x}$　　　　　　　　　　(as, $L \sin\Theta = \bar{x}$)

2.4 Centre of Pressure

As indicated earlier the intensity of pressure on an immersed surface is not uniform, but increases with depth. As the pressure is greater over the lower portion of the figure, the resultant pressure, on an immersed surface, will act at some point, below the centre of gravity of the immersed surface and towards the lower edge of the figure. The point through which this resultant pressure acts is known as 'centre of pressure' and is always expressed in terms of depth from the liquid surface.

2.4.1 Centre of pressure of a vertically immersed surface: Consider a plane surface immersed vertically in a liquid as shown in Figure (2-7):

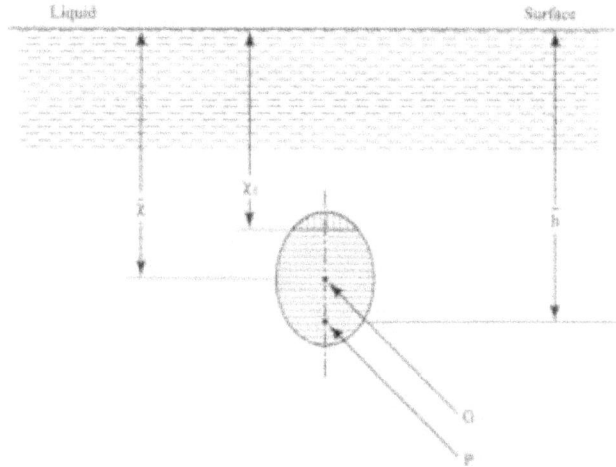

Fig (2-7) Surface immersed vertically in liquid

Dividing the whole surface into a number of small parallel strips as shown in Figure (2-7):
Let,
w = sp. weight of the liquid,
A = area of the surface,
$\bar{x}$ = depth of the centre of gravity of the surface from the liquid surface,
a_1, a_2, a_3 .... = areas of the strips, and
x_1, x_2, x_3 .... = depth of corresponding strips from the liquid surface.

Pressure on the first strip = w a_1 x_1
Moment of this pressure about the liquid surface = w a_1 x_1 x_1 = w a_1 x_1^2.

Similarly, moment of the pressure on the second strip about the liquid surface = w a_2 x_2^2.

And moment of the pressure on the third strip about the liquid surface = w a_3 x_3^2.

The sum of moments of all such pressures about the liquid surface,
$$M = w\,a_1\,x_1^2 + w\,a_2\,x_2^2 + w\,a_3\,x_3^2 + ... = w\,(a_1\,x_1^2 + a_2\,x_2^2 + a_3\,x_3^2\,...) = w\,I_o \qquad (i)$$
Where,
$I_o = (a_1\,x_1^2 + a_2\,x_2^2 + a_3\,x_3^2\,...)$ = Moment of inertia of the surface about the liquid surface
(also known as 'second moment of area')

As the sum of the moments of the pressure = $P \times \bar{h}$ $\qquad\qquad\qquad$ (ii)
Where,
P = total pressure on the surface,
$\bar{h}$ = depth of the centre of pressure from the liquid surface.

Equating (i) and (ii):
$P \times \bar{h} = w\,I_o$
$w\,a\,\bar{x}\,\bar{h} = w\,I_o$ $\qquad\qquad$ (as P = w a $\bar{x}$)
$\bar{h} = I_o / A\,\bar{x}$ $\qquad\qquad\qquad\qquad\qquad\qquad\qquad\qquad\qquad$ (iii)

Applying the theorem of parallel axis to get:
$I_o = I_g + Ah^2$
Where,
I_g = moment of inertia of the figure, about the horizontal axis through its centre of gravity,
h = distance between the liquid surface and the centre of gravity of the figure ($\bar{x}$ in this case)

Rearranging equation (iii) we get:
$$\bar{h} = (I_g + A\,\bar{x}^2) / A\,\bar{x} = (I_g / A\,\bar{x}) + \bar{x}$$
Thus the centre of pressure is always below the centre of gravity of the area by a distance =
$I_g / A\,\bar{x}$

TABLE (1)
Centre of gravity (G) and moment of inertia (I)
of some geometrical figures

NO	Name of Figure	C.G from the base	(I) about an axis passing through C.G and parallel to base	(I) about base
1		X = h/3	$bh^3/36$	$bh^3/12$
2		X = 2h/3	$bh^3/36$	$bh^3/12$
3		X = d/2	$bh^3/12$	$bh^3/3$
4		X= d/2	$\pi/64\,(d)^4$	-

2.4.2 Centre of pressure of an inclined immersed surface: Consider a plane inclined
surface immersed as shown in Figure (2-8):

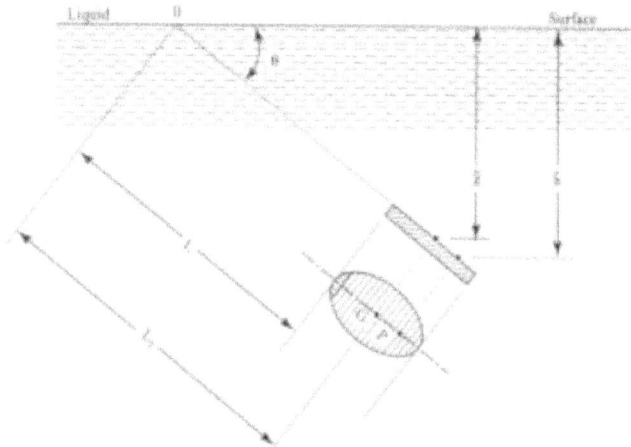

Figure (2-8) Centre of pressure for inclined surface

Divide the whole surface into a number of small parallel strips:
Let,
w = sp. weight of the liquid,
A = area of the surface,
$\bar{x}$ = depth of centre of gravity of the surface from the liquid surface
Θ = angle at which the immersed surface is inclined with the liquid surface
a_1, a_2, a_3 .... = areas of the strips, and
L_1, L_2, L_3 .... = distance of corresponding strips from 0.

Pressure on the first strip = $w\, a_1\, L_1\, \sin\Theta$
Moment of this pressure about 0 = $w\, a_1\, L_1\, \sin\Theta\, L_1 = w\, a_1\, L_1^2\, \sin\Theta$

Similarly,
moment of the second strip about 0 = $w\, a_2\, L_2^2\, \sin\Theta$
and moment of the third strip about 0 = $w\, a_3\, L_3^2\, \sin\Theta$

The sum of moment of all such pressures about 0:
$M = w\, a_1\, L_1^2\, \sin\Theta + w\, a_1\, L_1\, \sin\Theta + w\, a_3\, L_3^2\, \sin\Theta + ...$
     $= w\, \sin\Theta\, (a_1\, L_1^2 + a_1\, L_1 + a_3\, L_3^2 + ...)\, w\, \sin I_0$
$I_0 = (a_1\, L_1^2 + a_1\, L_1 + a_3\, L_3^2 + ...)$ = Moment of inertia of the surface about 0 (also known as second moment of area)

$M = w\, \sin\Theta\, I_0$ (i)

We know that the sum of the moments of all such pressures is also $= P\bar{h}/\sin\Theta$ (ii)
Where,
P = total pressure on the surface, and
$\bar{h}$ = depth of centre of pressure from the liquid surface

Equating (i) and (ii):
$P\bar{h}/\sin\Theta = w \sin\Theta I_0$
$w A \bar{x} \bar{h}/\sin\Theta = w \sin\Theta I_0$ (as $P = w A \bar{x}$)
$\bar{h} = I_0 \sin^2\Theta / A \bar{x}$ (iii)

From the theorem of parallel axis, we have:
$$I_0 = I_g + Ah^2$$
Where,
I_g = moment of inertia of the figure about horizontal axis through its centre of gravity,
h = distance between the liquid surface and the centre of gravity of the figure ($\bar{x}$ in this case).

Rearranging equation (iii), we get:
$$\bar{h} = (I_g \sin^2\Theta / A\bar{x}) + (A\bar{x}^2 \sin^2\Theta / A\bar{x} \sin^2\Theta)$$
$$= (I_g \sin^2\Theta / A\bar{x}) + \bar{x}$$
Thus the centre of pressure is always below the centre of gravity of the area by a distance = log $\sin^2\Theta$ / A$\bar{x}$.

2.4.3 Centre of pressure of an irregular section: The centre of pressure of an irregular section (i.e. a section with cut out hole or other irregular section) is obtained as follows:

(1) Split up the irregular section into convenient sections (i.e. rectangles, triangles or circles).
(2) Calculate the pressures: p_1, p_2 ... on all sections.
(3) Then calculate the total pressure 'p' on the whole section by the algebraic sum of the different pressures.
(4) Calculate the depths of centres of pressure $\bar{h}_1$, $\bar{h}_2$ ... for all the sections from the water surface.
(5) Equate $p\bar{h} = p_1\bar{h}_1 + p_2\bar{h}_2 + \dots\dots$

Where, $\bar{h}$ = depth of centre of pressure of the section from the water level.

2.5 Pressure on a Curved Immersed Surface

The total pressure on a curved surface, when immersed in a liquid, cannot be found out readily by the method explained earlier.

However, the same can be conveniently obtained by calculating the horizontal and vertical components of the resultant or total pressure, which is then combined together to give the total pressure on the curved surface.

Consider a curved surface AB immersed in a liquid. Let BC be the vertical projection and AC the horizontal project of the curved surface as shown in Figure (2-7):

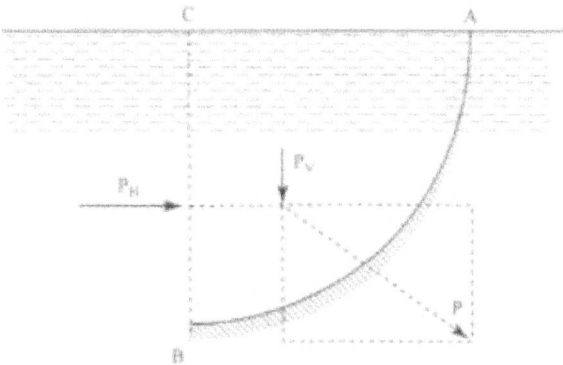

Fig (2-7) Curved surface immersed in liquid

The horizontal pressure P_H will be the total horizontal pressure on the projection BC of the curved surface and will act through the centre of pressure of the surface. The vertical pressure P_V will be the total weight of the liquid in the portion ABC and will act through the centre of gravity of the volume ABC.

The total pressure or resultant pressure may be found out by the relation:

$$P = \sqrt{P_H^2 + P_V^2}$$

The inclination of the resultant pressure is given by the equation:

$$\tan \alpha = P_V / P_H$$

Where α is the angle, which the resultant pressure makes with the horizontal.

2.6 Buoyancy

The tendency of a fluid to uplift a submerged body, because of the upward thrust of the fluid, is known as the force of buoyancy or simply buoyancy. It is always equal to the weight of the fluid displaced by the body. It may be noted that if the force of buoyancy is greater than the weight of the body, it will be pushed up till the weight of the fluid displaced is equal to the weight of the body. Hence, the body will float. On the other hand, if the force of buoyancy is less than the weight of the body, it will sink.

2.6.1 Centre of buoyancy: The centre of buoyancy is the point, through which the force of buoyancy is supposed to act. It is always the centre of gravity of the volume of the liquid displaced. In other words, the centre of buoyancy is the centre of the area of the immersed section.

2.6.2 Metacentre: Whenever a body, floating in a liquid, is given a small angular displacement, it starts oscillating about some point. This point, about which the body starts oscillating, is called 'metacentre':

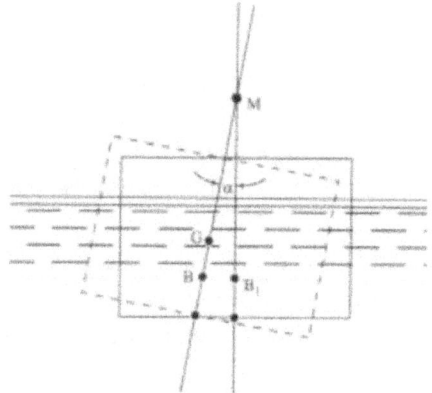

Fig (2-10) Metacentre

In other words, the metacentre may also be defined as the intersection of the line passing through the original centre of buoyancy and C.G of the body; and the vertical line through the new centre of buoyancy as shown in Figure (2-10).

2.6.3 Metacentric height: The distance between the centre of gravity, of a floating body, and the metacentre [i.e. distance GM as shown in Figure (2-10)] is called 'Metacentric Height'.

The metacentric height of a floating body may be found out by either of the following methods:
(1) Analytical determination of metacentric height, and
(2) Experimental determination of metacentric height.

2.6.4 Analytical determination of metacentric height:

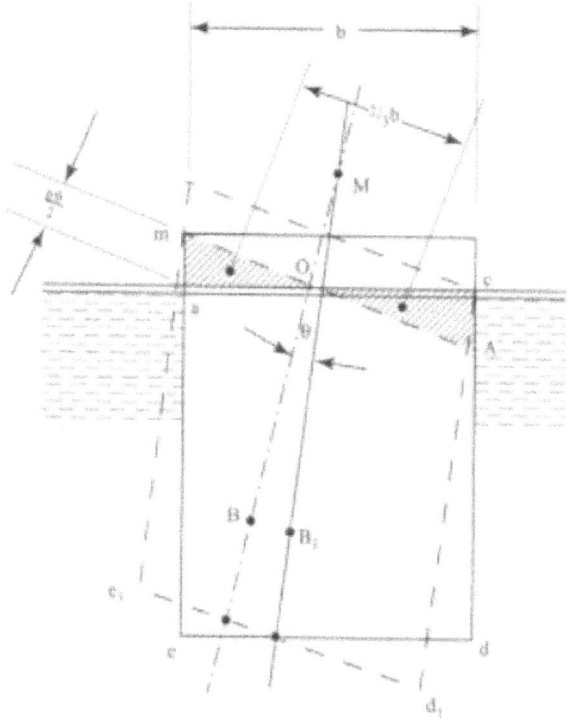

Fig (2-11) Metacentric height

Consider a ship floating freely in water. Let the ship be given a clockwise rotation through a small angle θ (in radians) as shown in Figure (2-11). The immersed section has now changed from a c d e to a c d_1 e_1.

The previous centre of buoyancy B has changed to B_1. It is to be noted that the triangular wedge 'a o m' is out of water, whereas the triangle 'o c n' is under water. Since the volume of water displaced remains the same, therefore, the two triangular wedge must have equal areas.

A little consideration will show that as the triangular wedge 'a o m' has come out of water, thus decreasing the force of buoyancy on the left, therefore, it tends to rotate the vessel in an anti-clockwise direction. Similarly, as the triangular wedge 'o c n' is underwater, thus increasing the force of buoyancy on the right side, and tends to rotate the vessel in an anti-clockwise direction.

Let,
L = length of the ship,
b = breadth of the ship,
θ = very small angle (in radians) through which the ship is rotated about O, and

v = volume of water displaced by the ship.
Therefore,
a m = b Θ / 2

Volume of wedge of water a o m = ½ (b/2 x a m) L = ½ (b/2 x bΘ/2)L   (as a m = bΘ/2)
$$= b^2\Theta L/8 \qquad\qquad (i)$$

Weight of wedge of water = w $b^2\Theta L/8$
Where,
w = Sp. weight of water

Similarly,
weight of wedge o c n of water = (w $b^2\Theta L$) / 8

Equating the amount of weights of wedges a o m and c o n about M and equating the same to the moment of the weight of liquid displaced we get:

$$(w\ b^2\Theta L/8 \times b/3) + (w\ b^2\Theta L/8 \times b/3) = w\ v \times BB_1$$
$$w\ b^3\Theta L/12 = w\ v \times BB_1 \qquad\qquad (ii)$$

Substituting the value of $Lb^3/12 = I$ i.e. moment of inertia of the plan of the ship; and
$BB_1 = BM \times \Theta$ in equation (ii), we get:
$$w\ I\ \Theta = w \times v\ (BM \times \Theta)$$
 Therefore,
$$BM = I/v$$
And metacentric height:
$$GM = BM \pm BG$$

(Note: Positive sign is to be used if G is lower than B and negative sign to be used if G is higher than B.)

2.6.5 Experimental determination of metacentric height: The metacentric height of a floating body, like a ship, may also be found out experimentally; provided the centre of gravity of the floating body is known.

Let all goods on ship to be arranged in such a way, that the ship is perfectly horizontal as shown in Figure 92-12):

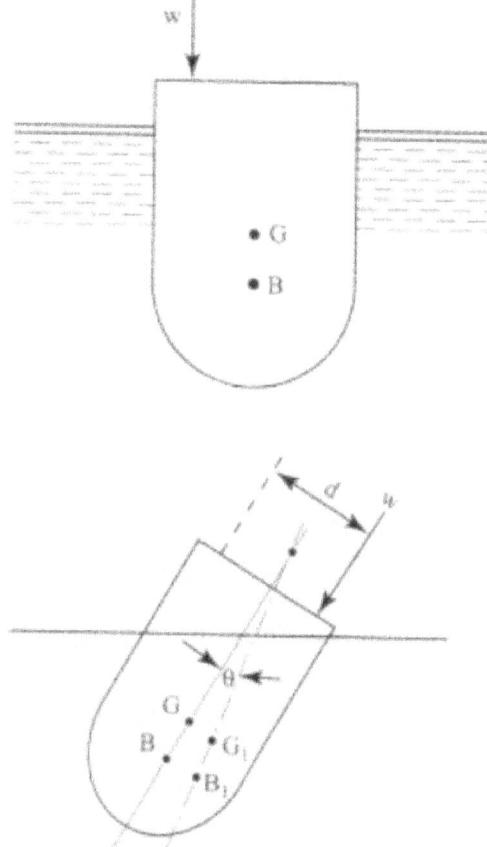

Fig (2-12) Determination of metacentre height experimentally

Let,

W = weight of the ship, and

G = centre of gravity of the ship.

Let a movable weight 'w' be moved right across the ship through a distance 'd' as shown in Figure (2-12).

Due to this movement of load 'w', the boat will tilt. Let this angle of title be Θ. Let the centre of gravity (G) move to a new position G_1 and the centre of buoyancy (B) to a new positionB_1. Joining B_1 and G_1 and extending the line upwards to meet the extension of the line passing through B and G, to meet at point M, which is the metacentre and GM is the metacentric height.

The effect of moving the load w to the right, through a distance d, will cause a clockwise couple, whose moment = w d   (i)

The weight of the boat W and the force of buoyancy will form an anticlockwise couple, whose moment = W G M tanΘ   (ii)

Since these two moments are equal, but opposite in directions, therefore equating (i) and (ii) we get:

$$W G M tanΘ = w d$$

Or,

$$G M = w d / W tanΘ$$

Where,
G M is the metacentric height.

2.7 The Hydrostatic Equation

The hydrostatic equation describes the behaviour of fluids in a gravitational field. It states that whenever there is no vertical motion, the difference in pressure (dp) between two levels (dz) is caused by the weight of the layer of fluid.

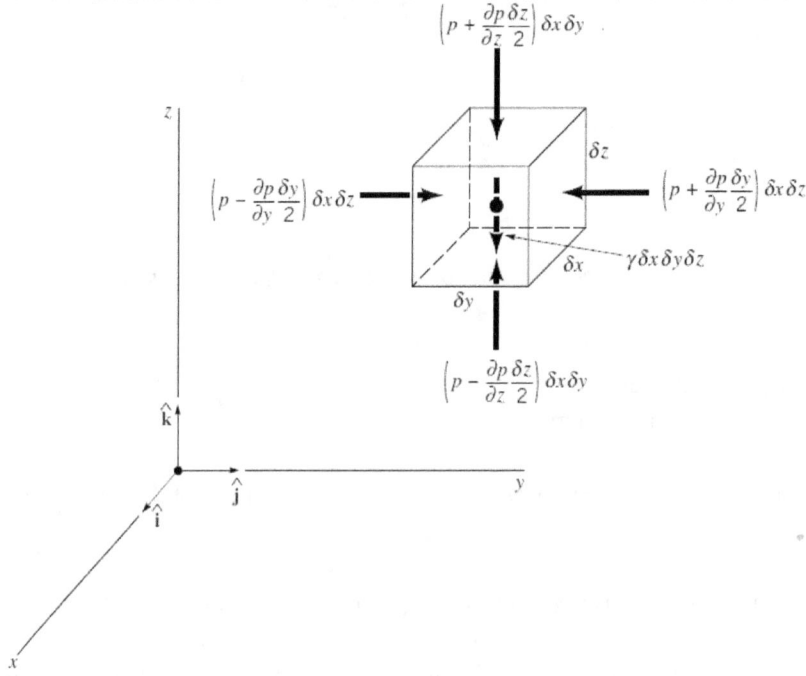

Fig (2-13) Balance of forces on element of fluid

The two acting forces taken into consideration are:

(a) Surface forces: These represent the pressure forces acting perpendicular to the surface, and

(b) Body forces: Is the force due to the weight force of the fluid within the surface element.

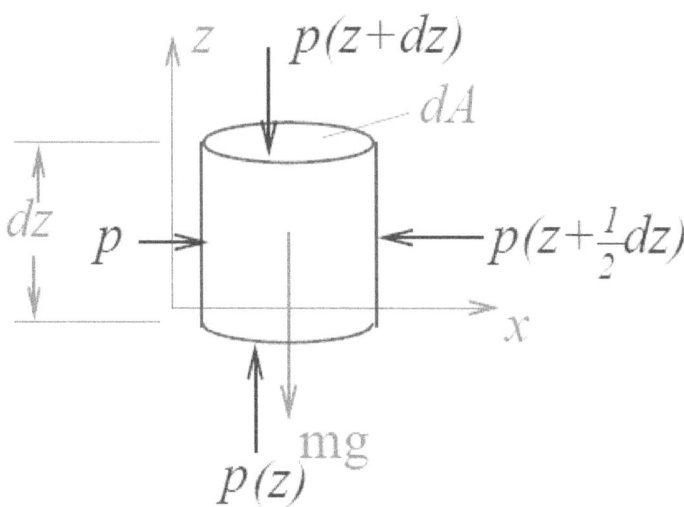

Fig (2-14) Forces on element of fluid

As the side to-side pressure forces, p (z + ½ dz), are equal and opposite, while the top and bottom forces acting on area dA are not equal, we get:

$$mg - p(z) dA + p(z + dz) dA = 0$$
$$(p\, dz\, dA)\, g - p(z) dA + p(z + dz) dA = 0$$
$$(\rho\, dz)\, g - p(z) + p(z + dz) = 0$$

Let, $\gamma = \rho\, g$,

$$\gamma\, dz - p(z) + p(z + dz) = 0$$
$$p(z + dz) = p(z) - \gamma\, dz \qquad (i)$$

Consider forces on a plug of water as shown in Figure (2-14), and using a Taylor series for vertical pressure variation we get:

$$p(z + dz) = p(z) + \partial p / \partial z\, dz + ... \qquad (ii)$$

By equating (i) and (ii) we get:

$$\partial p / \partial z = -\gamma \qquad (2\text{-}2)$$
$$\partial p / \partial x = 0 \qquad (2\text{-}3)$$

Equations (2-2) and (2-3) describe the variation of the pressure with respect to the horizontal and vertical displacement.

With liquids any density variation with pressure can be ignored in most engineering applications involving incompressible fluids:

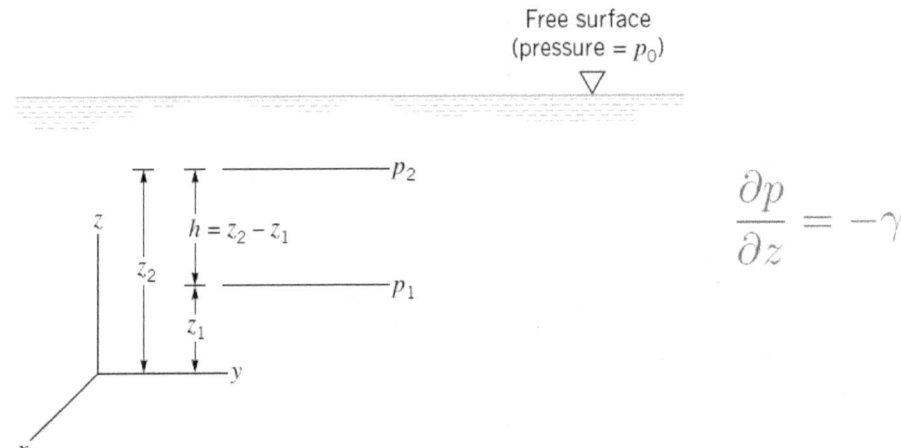

Fig (2-15) The pressure increases as the height decreases

With no x, y dependence, $\partial z \rightarrow dz$, hence:

$$dp/dz = -\gamma$$

$$p\,(z) = -\int_{z_1}^{z_2} \gamma\, dz + C$$

$$p\,(z) = (z_1 - z)\,\gamma + p_1 \qquad \text{as, } p\,(z_1) = p_1$$

We note that, the pressure increases as z decreases because it has to hold-up more of the fluid above.

2.8 Two-phase Horizontal Flow

The respective distribution of liquid and vapour phases in two-phase flow pipe is an important aspect of their description. Analogous to predicting the transition from laminar to turbulent flow in single phase flows, two phase flow pattern maps are used for predicting the transition from one type of two-phase flow to another.

There are two possibilities for two different materials to flow. The materials can flow in the same direction and it is referred as 'co-current flow'. When the materials flow in opposite directions, it is referred as 'counter-current'. In general, the co-current is the more common.

Generally, the counter-current flow has a limited length. In co-current flow, two liquids can have three main categories: vertical, horizontal, and in between.

The main complicating feature in horizontal flow regime is the gravitational forces acting on the liquid phase causing it to be displaced towards the bottom of the pipe. Consider a mixture that is almost completely liquid; this means that the ratio of: Liquid velocity / Gas velocity (V_L/V_G) is large. In this case, the gas is present in the form of small bubbles that will rise to the top portion of the pipe. As the gas flow rate increases, the bubbles become larger, and further increase of V_G leads to what we call annular mist flow through the intermediate stage of slug flow.

Figure (2-16) below shows plots of typical flow regimes for horizontal gas-liquid flows in a pipe:

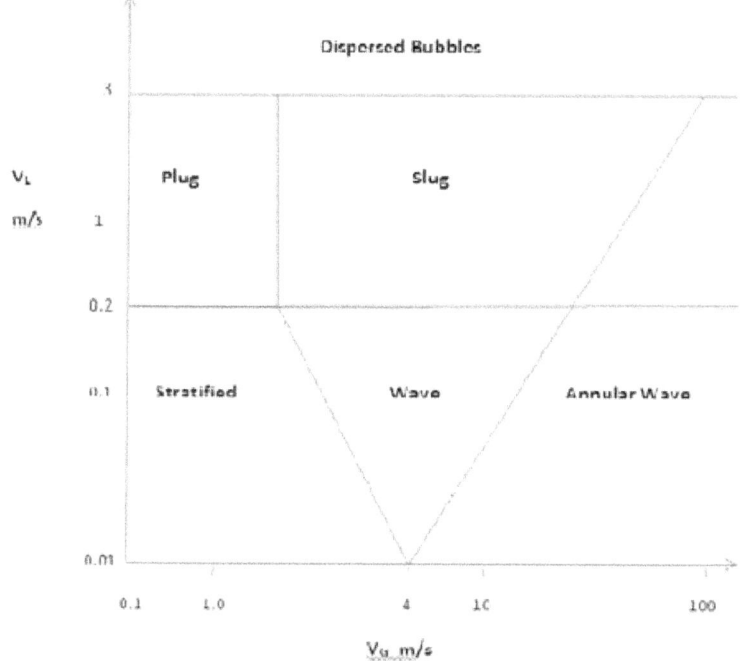

Fig (2-16) Plots of typical flow regimes for horizontal gas-liquid flows

The various flow regimes are dependent on v_G and v_L:

1) Dispersed bubble flow:

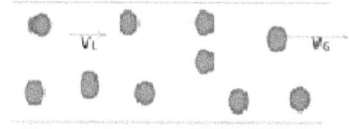

Diagram (1)
The bubbles tend to flow at the top of the pipe.

2) Plug flow - elongated bubble:

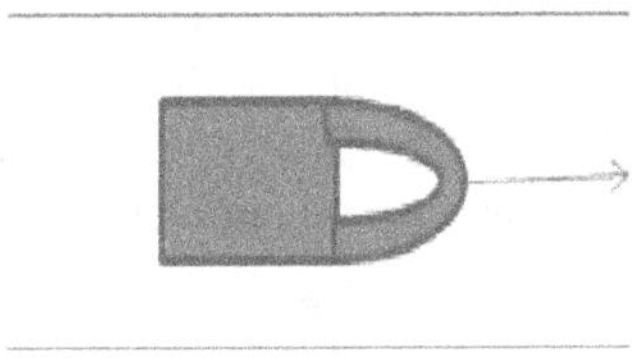

Diagram (2)

Bullet shape bubbles occur, but they tend to move along in a position closer to the top of the pipe.

3) Slug flow:

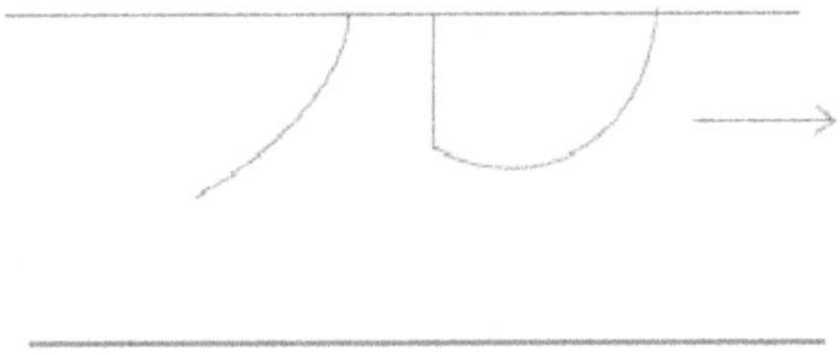

Diagram (3)

Gravitational effect with two layers one liquid at the lower part of the pipe and gas along the top part.

4) Stratified flow:

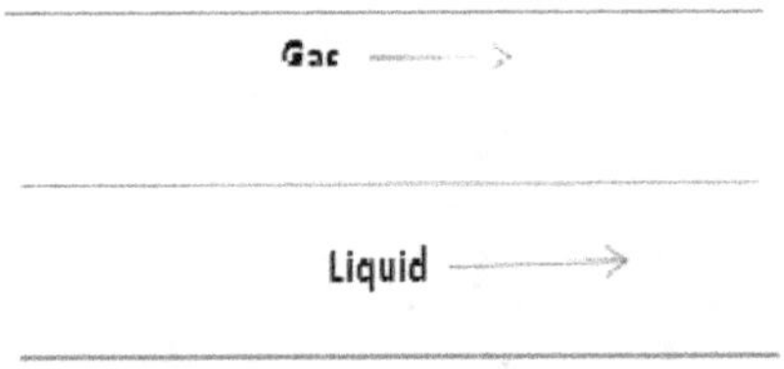

Gas ———— >

Liquid ————⟶

Diagram (4)

As the gas velocity is increased in stratified flow on the gas-liquid interface giving two distinct layers.

5) Wave flow:

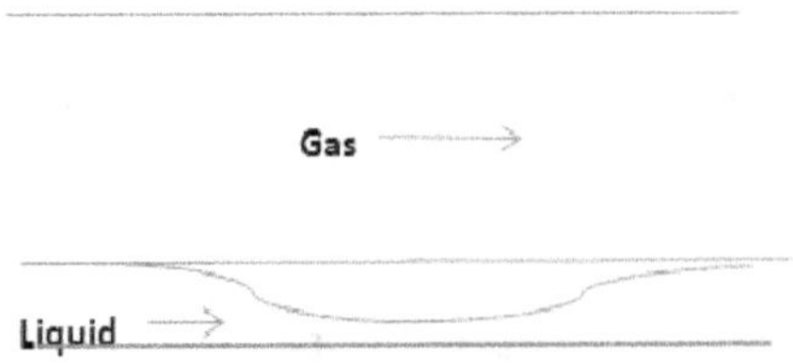

Gas ———⟶

Liquid —⟶

Diagram (5)

As the gas velocity is further increased, the wavy flow region eventually becomes high to reach the top of the pipe.

6) Annular flow:

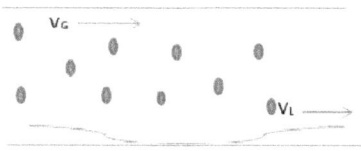

Diagram (6)

The liquid flows on the wall of the pipe as a film and the gas phase flows in the centre. Usually, some liquid phase is entrained as small droplets in gas core.

The modelling of gas-liquid flow in horizontal pipes is also of considerable importance in reactor safety analysis. For example, in case of a breakage of the main cooling circuit of a pressurized water reactor, the pressure losses of the gas-liquid flow governs significantly the loss of coolant rate. The pressure losses depend on the flow regime which is largely determined by the steam mass fraction and the superficial velocities, but will also reflect the pipe topology. An elbow and a transition from vertical to horizontal two-phase flow may lead to stratified flow due to the centrifugal separation in a certain part of the horizontal pipe, where the above variables would indicate transition to slug flow.

2.8.1 Pressure drop in two-phase flow: Two-phase flow is difficult subject principally because of the complexity of the form in which the two fluids exist inside the pipe. Pressure drop in two-phase flow is a major design variable, governing the pumping power required to transport two-phase fluids. The orientation of the pipe makes a difference in the flow regime because of the role played by gravity and the density difference between the two fluids.

The usual question for the engineer is that of calculating the pressure drop required to achieve specific flow rates of the gas and the liquid through a pipe of a given diameter.

The simplest approach to the prediction of two-phase flows is to assume that the phases are thoroughly mixed and can be treated as a single-phase flow. For homogenous model, the pressure (dp/dz) is given by:

$$- dp/dz = \tau_0 B/A + d\ (\dot{M}^2/\rho_H)/dz + g\ \rho_H \sin\alpha \qquad (2\text{-}4)$$

Where,
τ_0 is the wall shear stress,
B the pipe periphery,
A the pipe cross-sectional area,
$\dot{M}$ the mass flow,
z the axial distance,
g gravitational acceleration,
α the angle of inclination of the pipe to the horizontal, and

ρ_H the homogeneous density given by:

$$\rho_H = \rho_G \rho_L / x \rho_L + (1 - x) \rho_G \qquad (2\text{-}5)$$

where,

x is the quality (fraction of the total flow which is vapour).

The three terms on the right-hand side of equation (2-4) may be regarded respectively as the frictional pressure gradient ($- dp_L/dz$), the acceleration pressure gradient ($- dp_a/dz$) and the gravitational gradient ($- dp_g/dz$). Thus:

$$- dp/dz = - dp_L/dz - dp_a/dz - dp_g/dz \qquad (2\text{-}6)$$

The frictional pressure gradient term in the homogeneous model is often related to a two-phase friction factor f_{TP} as follows:

$$- dp_L/dz = \tau_0 B/A = 2 \, f_{TP} \, \dot{M}^2/D \, \rho_H \qquad (2\text{-}7)$$

Where,

D   is the pipe diameter, and
f_{TP} is related (within the normal single-phase friction factor relation) to a two-phase Reynolds number defined as:

$$Re_{TP} = \dot{M} D/\mu_{TP} \qquad (2\text{-}8)$$

Where,
μ_{TP} is a two-phase viscosity.
μ_{TP} is expressed by the relation:

$$1/ \mu_{TP} = x/\mu_G + (1 - x)/\mu_L \qquad (2\text{-}9)$$

Where,
μ_G and μ_L are the gas and liquid viscosities respectively.

2.9 Engineering Application of Hydrostatics

Strictly in the engineering field, the hydrostatic pressure is either utilised in the working of a hydraulic structure, or a structure is checked to withstand the hydraulic pressure exerted on it. Tus the study of the subject hydrostatics is of much importance while designing all sorts of hydraulic structures or hydraulic devices. Presently, we shall discuss the practical application of the hydraulics on the following structures:

1) Sluice gates,
2) lock gates,
3) Masonry walls, and
4) Dams.

2.9.1 Water pressure on sluice gates: A sluice gate consists of two vertical plates, known as skin plate. In between these two skin plates, a number of I-beams are provided horizontally to withstand the water pressure. As the water pressure varies with depth, therefore the spacing between the I-beams is more of the bottom than at the top of the sluice gate. A sluice gate is provided, in the path of a river or stream, to regulate the flow of water. Water acts on both sides of sluice gate and the resultant pressure is obtained as follows:

Let,
P_1 = pressure on upstream side of the gate,
A_1 = wetted area on the upstream side of the gate, and
$\bar{x}_1$ = depth of centre of gravity of the wetted area on the upstream side of the gate.

Therefore,
$$P_1 = w\,A_1\,\bar{x}_1 \qquad\qquad (i)$$
Similarly, let
P_2 = pressure on downstream side of the gate,
A_2 = wetted area on the downstream side of the gate, and
$\bar{x}_2$ = depth of centre of gravity of the wetted area on the downstream side of the gate.

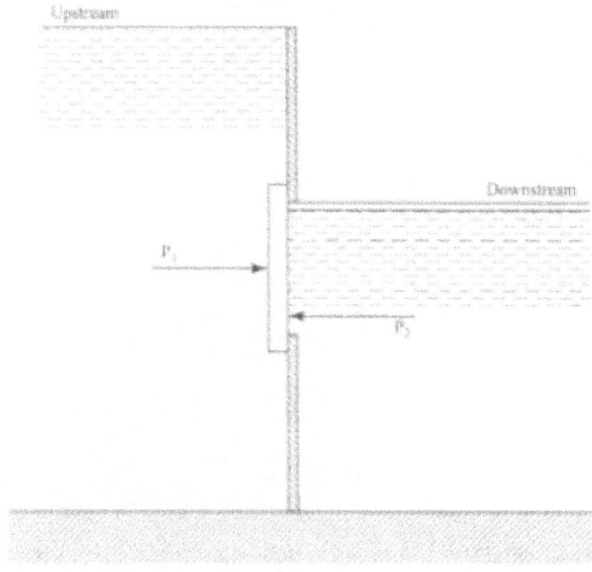

Fig (2-17) Sluice gate

Therefore,
$$P_2 = w\,A_2\,\bar{x}_2 \qquad\qquad (ii)$$

As the two pressures are acting in opposite directions, therefore the resultant pressure is:

$$P = P_1 - P_2$$

The point of application of the resultant pressure (P) may be found out by taking moments of the two pressures (P_1 and P_2) about the top or bottom of the gate.

2.9.2 Water pressure on lock gates: When a dam is constructed across a river or a canal, the water levels on both sides of the dam will be different. If navigation or boating is required, then a chamber known as a lock is constructed between these two different water levels.

Two sets of lock gates (one on the upstream side and the other on the downstream side) are provided as shown in Figure (2-18):

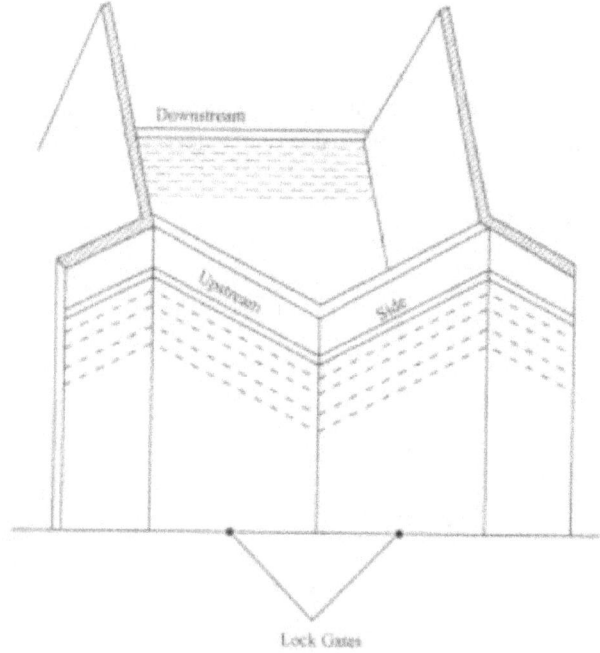

Fig (2-18) Lock gate

In order to transfer a boat from the upstream (i.e. from a higher water level) to the downstream (i.e. to a lower water level) the upstream gates are opened (while the downstream gates are closed) and water level in the chamber rises up to the upstream water level. The boat is then admitted in the chamber. Then the upstream gates are closed and the downstream gates are opened and the water level in the chamber is lowered up to the downstream water level. Now the boat can proceed further downwards. If the boat is to be transferred from the downstream to upstream side, the above procedure is reversed.

Now consider a set of lock gates AB and BC hinged at the top and bottom A and C respectively as shown in Figure (2-19):

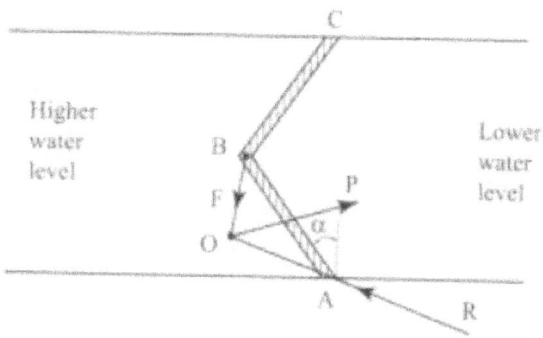

Fig (2-19) Lock gates hinged at the top and bottom

These gates will be held in contact at B by the water pressure, the water level being higher on the left-hand side of the gates:

Let,
P = water pressure on the gate AB acting at right angles on it,
F = force exerted by the gate BC acting normally to the contact surface of the two gates AB and BC (also known as reaction between the two gates), and
R = Reaction at the upper and lower hinge.

Since the gate AB is in equilibrium under the action of the above three forces, therefore they will meet at one point. Let P and F meet at O, then R must pass through this point.

Let,
α = inclination of the lock gate with the normal to the side of the lock.

From the geometry of figure ABO, we find that it is an isosceles triangle having its angle OBA and OAB both equal to α.

Resolving the forces, at o, parallel to AB,
$$R \cos\alpha = F \cos\alpha$$
Therefore,
$$R = F \qquad\qquad (i)$$

Now resolving the forces at right angles to AB:

$$P = R \sin\alpha + F \sin\alpha = 2 R \sin\alpha \qquad (as\ R = F)$$

Therefore,
$$R = P/2 \sin\alpha$$
Or,
$$F = P/2 \sin\alpha$$

Considering the water pressure on hinges of the gates:
Let,
H_1 = height of water to the left of the gate,
H_2 = height of water to the right of the gate,
P_1 = total pressure of water on the left of the gate,
P_2 = total pressure of water on the right of the gate,
R_T= reaction of top hinge,
R_B= reaction of bottom hinge,
A_1= wetted area of one gate on left of the gate, and
A_2= wetted area of one gate on right of the gate.

Since the total reaction (R) will be shared by two hinges (R_T and R_B),
$$R_T + R_B = R \qquad\qquad (ii)$$

Knowing that the total pressure on the lock gate as:

$$P = w\,A\,\bar{x}$$
Or,
$$P_1 = w\,A_1\,(H_1/2) \qquad\qquad (as\ x_1 = H_1/2)$$
Similarly,
$$P_2 = w\,A_2\,(H_2/2)$$

Since the direction of P_1 and P_2 are in opposite directions, therefore the resultant pressure will be:
$$P = P_1 - P_2$$

As the P_1 will act through its centre of pressure, which is at a height of ($H_1/3$) from the bottom of the gate. Similarly, pressure P_2 will also act through its centre of pressure, which is at a height of ($H_2/3$) from the bottom of the gate.

A little consideration will show that half of the resultant pressure (i.e. P_1 - P_2 or P) will be resisted by the hinges of one lock gate (as the other half will be resisted by the hinges of the other lock gates).

Taking moments about the lower hinge:
$$R_T \sin\alpha \times h\ (P_1/2 \times H_1/3) - (P_2/2 \times H_2/3) \qquad\qquad (iii)$$

Where (h) is the distance between the two hinges.

Also resolving the forces horizontally,

$$P_1 - P_2 = R_B \sin\alpha + R_T \sin\alpha \qquad\qquad (iv)$$

From equations (iii) and (iv) the values of R_B and R_T may be found out.

If the two hinges are not located at the extreme top and bottom of a lock gate, but are located at some distance from the top and bottom then the hinge reactions may be found out as discussed below:
Let,
d = distance of the bottom hinge from the bottom of the gate.

Taking moments about the bottom hinge we get:

$$R_T \sin\alpha\, h = [P_1/2 \times (H_1/3 - d)] - [P_2/2 \times (H_2/3 - d)]$$

2.9.3 Water pressure on masonry walls: Consider a vertical masonry wall having water on one of the sides as shown in Figure (2-20):

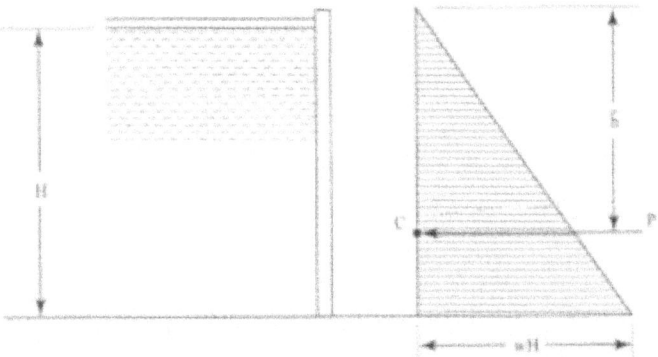

Fig (2-20) Masonry wall

Consider a unit length of the wall. Knowing that the pressure will act perpendicular to the wall, and with little consideration will show that the intensity of pressure at water level will be zero and will increase by a straight line law to (w h) at the bottom, as shown in Figure (2-20). Tus, the pressure diagram will be a triangle.

The total pressure on the wall will be the area of the triangle i.e.

$$P = w\, H/2 \times H = w\, H^2/2$$

This pressure will act through the centre of gravity of the pressure diagram.
Let,
$\bar{h}$ = depth of the centre of pressure from the water surface.

Knowing that the centre of gravity (C.G) of a triangle is at a height of H/3 from the base;
where,
H is the height of the triangle.
Therefore,

$$\bar{h} = H - H/3 = 2H/3$$

Thus the pressure of water on a vertical wall will act through a point at a distance H/3 from the bottom;
where,
H is the depth of water.

2.9.4 Water pressure on masonry dams: Dams are constructed in order to store large quantities of water, for the purpose of irrigation and power generation. A dam maybe of any cross-section, but the following are important from the subject point of view:

(1)Rectangular dams, and
(2) Trapezoidal dams.

Water pressure on rectangular dams:
Consider a rectangular dam having water on one of its sides, as shown on Figure (2-21).

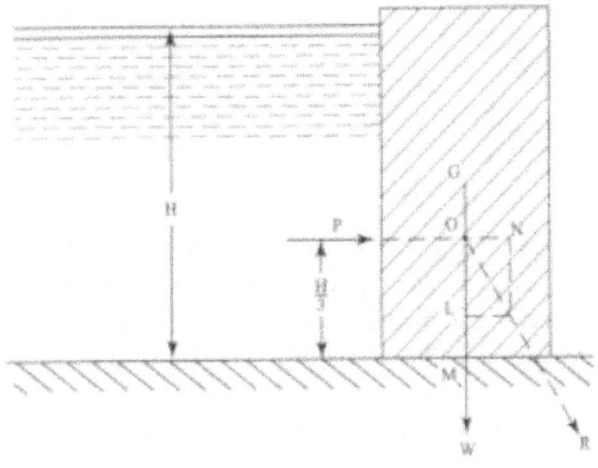

Fig (2-21) Rectangular dam

Now consider a unit length of the dam:

Let,

H = height of water stored by the dam,

P = total pressure of water = w $H^2/2$

We know that the pressure P will act as height of H/3 above the base of the dam. Let w be the weight of masonry per unit length of the dam. We know that w will act downwards through the centre of gravity of the dam section.

Hence, the resultant pressure P and w will be given by the relation,

$$R = \sqrt{P^2 + w^2}$$

And the inclination of the resultant with the vertical, Ѳ, is given by the relation:

$$\tan Ѳ = P/w$$

Let x be the horizontal distance between the centre of gravity of the dam and the point through which the resultant cuts the base. It may be found from the similar triangles OLR and OMR, i.e.

$$MR/OM = ON/OL$$
$$x / (H/3) = P/w$$

Or, $x = P/w \times H/3$

2.9.5 Water pressure on trapezoidal dams: A trapezoidal dam is more economical and also easier to construct than a rectangular dam.

Consider a trapezoidal dam having water on one of its sides (say vertical side) as shown in Figure (2-22):

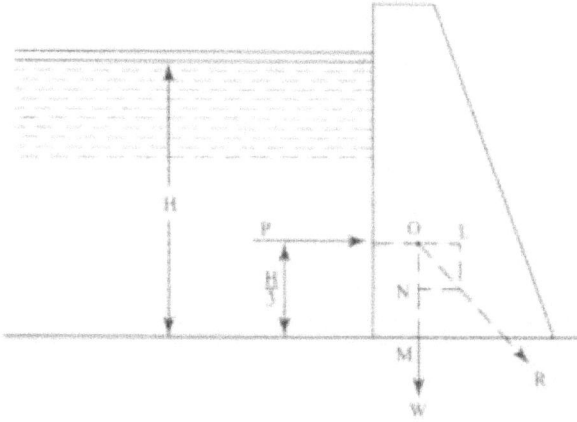

Fig (2-22) Trapezoidal dam

Like the rectangular dam, the total pressure on a trapezoidal dam will be given by the relation,

$$P = w\,H^2/2$$

And the horizontal distance between the centre of gravity of the dam and the point, at which the resultant cuts the base, is also given by the relation,

$$x = (P/w) \times (H/3)$$

2.9.6 Flow around bridge piers and through waterways:

a) Bridge piers:

Single-span bridge have abutment at each end that support the weight of the bridge and secure as retaining walls to reset lateral movement of the earth fill of the bridge approach.

Fig (2-23) 3 rows of piers with arches springing from them support of the bridge

Multi-span bridges require piers to support the end spans between these abutments.

b) Waterways:

Modified natural waterways and artificial drains, include timbered drains, concrete-lined channels, concrete dish channels and pipes. Vary in condition, with some having good base flows. It offers naturalisation opportunities:

Figure (2-24) Waterway

Environmental asset waterways include both natural waterways and some artificial drains, which often have good base flow. Unlike the downstream rivers which are often alongside public land, the tributaries are located mostly on private land.

2.9.7 Culverts: A culvert is a structure that allows water to flow under a road, railroad, trail, or similar obstruction. Typically embedded so as to be surrounded by soil, as shown in Figure (2-25 a & b):

Fig (2-25 a) Steel culvert with a plunge pool below

Fig (2-25 b) An Italian multiple culvert

Culverts may be used to form a bridge like structure to carry traffic. They come in many sizes and shapes including round, elliptical, flat-bottomed, pear-shaped, and box-like constructions. Culverts may be made of concrete, galvanized steel, aluminium, or plastic, typically high density polyethylene. Two or more materials may be combined to form composite structures.

Check Your Knowledge

1. State and prove Pascal's law.
2. Distinguish between gauge pressure and absolute pressure.
3. Explain the term centre of buoyancy.
4. State the position of the centre of buoyancy of a floating body.
5. Define the term metacentre.
6. Explain what is meant by metacentric height of a floating body.
7. Deduce an expression for the metacentric height of a floating body.
8. What are the conditions of equilibrium of a floating body?
9. Explain the stability of a floating body, with reference to its metacentric height. *Give neat sketch.*
10. Select and write the correct statement, giving reasons whenever necessary:
A floating body in a liquid is in equilibrium:

(i) When its centre of gravity is below its centre of buoyancy.

(ii) When its metacentric height is zero.

(iii) When its metacentre is below the centre of buoyancy.

(iv) When the metacentre is above the centre of gravity.

(v) None of these.

11. Illustrate the typical flow regimes for horizontal gas-liquid flows.

3. Pressure Measurement

3.1 Measurement of Fluid Pressure

The principles, on which all the pressure-measuring devices are based are almost the same. However for convenience we can split them into the following two types:

(1) By balancing the liquid column (whose pressure is to be found out) by the same or another column. These are also called tube gauges to measure the pressure.

(2) By balancing the liquid column (whose pressure is to be found out) by a spring or dead weight. These are also called mechanical gauges to measure the pressure.

3.1.1 Tube gauges to measure the fluid pressure: The device used for measuring the fluid pressure, by tube gauges are:

(1) Piezometer tube, and
(2) Manometer

Piezometer tube: A piezometer tube is the simplest form of manometer, used for measuring moderate pressures. It consists of a tube which is open at one end to the atmosphere, in which the liquid can rise freely without overflow. The height, to which the liquid rises up in the tube gives the pressure head directly.

If the pressure of a liquid flowing in a pipe is to be found out, the piezometer tube is connected to the pipe. While connecting the piezometer to a pipe, care should be taken so that the tube should not project inside the pipe beyond the surface. All burrs and roughness near the hole must be removed and the edge of the hole should be rounded off.

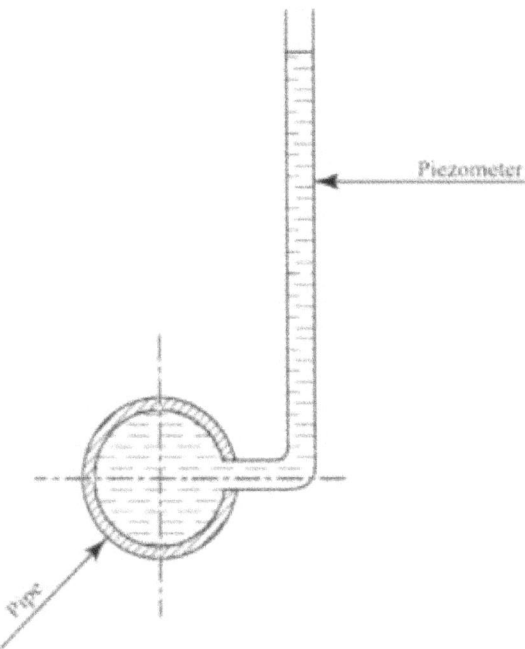

Fig (3-1) Piezometer tube

It may be noted that the piezometer tube is meant for measuring gauge pressure only as the surface of the liquid, in the tube, is exposed to the atmosphere. A piezometer tube is also not suitable for measuring negative pressure, as in such cases the air will enter in the pipe through the tube.

3.2 Manometer: A manometer is an improved type of piezometer tube. With the help of a manometer, we can measure comparatively high pressures and negative pressures. The following are a few types of manometers:

(1) Simple manometer,
(2) Micro-manometer,
(3) Differential manometer, and
(4) Inverted differential manometer.

3.2.1 Simple manometer: As seen earlier a piezometer is not suitable for the measurement of high pressures or negative pressures. The piezometer tube was then improved in such a way making it possible to measure high as well as negative pressures. This improved piezometer is known as a manometer.

A simple manometer, in the simplest form, is a tube bent in U-shape; one end of which is attached to the gauge point and the other is open to the atmosphere.

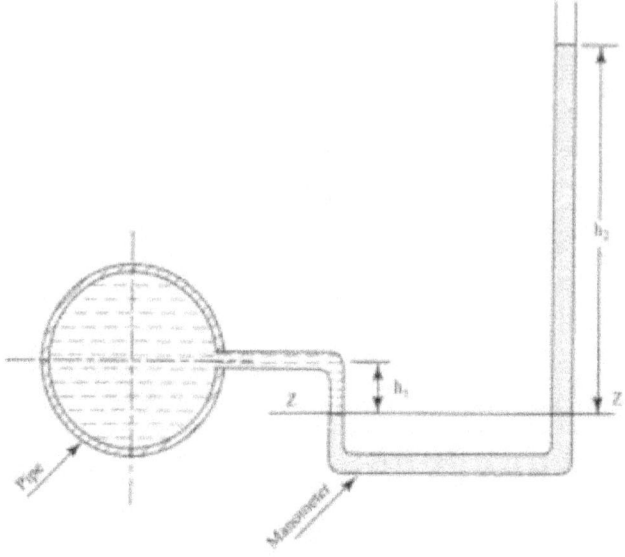

Fig (3-2) Manometer

The liquid used in the tube or in the simple manometer is generally mercury, which is 13.6 times heavier than water. Hence, it is also suitable for measuring high pressure.

Now consider a simple manometer connected to a pipe containing a light liquid under a high pressure. The high pressure in the pipe will force the heavy liquid, in the left side of the U-tube, to move downward. This downward movement of the heavy liquid in the left will cause a corresponding rise, of the heavy liquid, in the right side. The horizontal surface at which the heavy and light liquid meets in the left side is known as a common surface or datum line.

Let z-z be the datum line,
Let,
h_1 = height of the light liquid in the left side above the common surface (mm),
h_2 = height of the heavy liquid in the right side above the common surface (mm),
h = pressure in the pipe, expressed in terms of head of water (mm),
s_1 = specific weight of the light liquid, and
s_2 = specific weight of the heavy liquid.

Since the pressures in the left side and right side above the datum line are equal we get:
Pressure in the left side above the datum line = $h + s_1 h_1$ (mm of water), and
Pressure in the right side above the datum line = $s_2 h_2$ (mm of water).
Equating these two pressures:

$$h + s_1 h_1 = s_2 h_2$$

$$h = (s_2 - h_2 - s_1 h_1) \text{ mm of water}$$

If a negative pressure is to be measured by a simple manometer, the same can be measured easily:

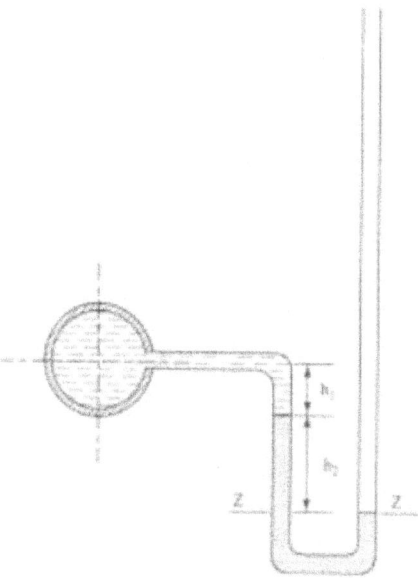

Fig (3-3) Manometer measuring negative pressure

In this case, the negative pressure in the pipe will suck the light liquid which will pull up the heavy liquid in the left side of the U-tube. Tis upward movement of the heavy liquid, in the left side, will cause a corresponding fall of the liquid in the right side.

In this case the datum z-z may be considered to correspond with the top level of the heavy liquid in the right column as shown in Figure (3-3).

Now the pressure in the left side above the datum line $= h + s_1 h_1 + s_2 h_2$, and pressure in the right side $= 0$.

Equating these two pressures:

$$h + s_1 h_1 + s_2 h_2 = 0$$
$$h = - s_1 h_1 - s_2 h_2 = - (s_2 h_2 + s_1 h_1) \text{ mm of water}$$

3.2.2 Micro-manometer: It is a modified form of manometer, in which cross sectional area of one of the sides (i.e. left side) is made much larger (about 100 times) than that of the other side. A micro-manometer is used for measuring small pressures; where accuracy is of much importance.

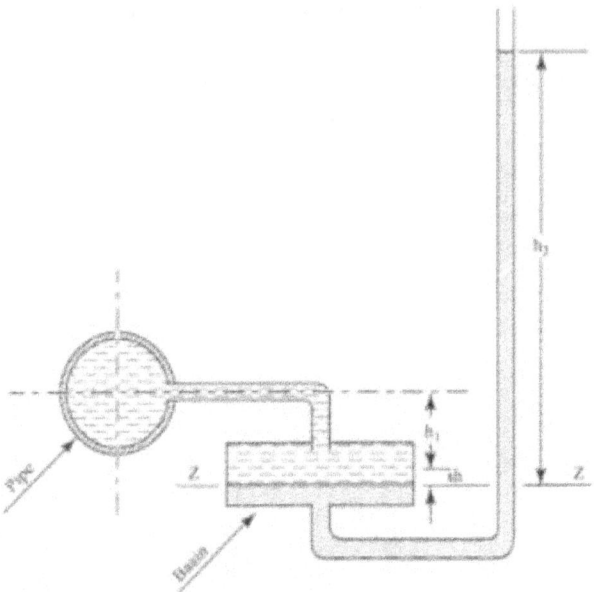

Fig (3-4) Micro-manometer

Now consider a micro-manometer connected to a pipe containing light liquid under very high pressure. The high pressure in a pipe will force the light liquid to push the heavy liquid in the basin downwards. Due to the large area of the basin the fall of heavy liquid level will be very small. This downward movement of the heavy liquid in the basin will cause a considerable rise of the heavy liquid in the right side.

Considering our datum line Z-Z corresponding to heavy liquid level before the experiment.
Let,
δh = fall of heavy liquid level in the basin (mm),
h_1 = height of light liquid above the datum line (mm),
h_2 = height of heavy liquid (after experiment) in the right side above the datum line (mm),
h = pressure in the pipe, expressed in terms of head of water (mm),
A = cross sectional area of the basin (mm^2),
a = cross sectional area of the tube (mm^2),
s_1 = specific weight of the light liquid, and
s_2 = specific weight of the heavy liquid.

As the fall of heavy liquid level in the basin will cause a corresponding rise of heavy liquid level:

$$A\ \delta h = a\ h_2 \quad \text{or} \quad \delta h = a/A\ h_2 \qquad \text{(i)}$$

Examining the horizontal surface in the basin at which the heavy and light liquid meets as datum line. Knowing that the pressure in the left side and right side, above the datum line we get:

Pressure in the left side above the datum line $= h + s_1 h_1 + s_1\,\delta h$ (ii)

And pressure in the right side above the datum line $= s_2 h_2 + s_2\,\delta h$ (iii)

Equating pressures in (ii) and (iii):
$$h + s_1 h_1 + s_1\,\delta h = s_2 h_2 + s_2\,\delta h$$
or,
$$h = s_2 h_2 + s_2\,\delta h - s_1 h_1 - s_2\,\delta h = s_2 h_2 - s_1 h_1 + \delta h\,(s_2 - s_1)$$

Substituting the value of δh from equation (i):

$$h = s_2 h_2 - s_1 h_1 + a/A\; h_2\,(s_2 - s_1)$$ (iv)

Sometimes the cross sectional area of the basin A is made very large and that of the tube, a, is made very small. Then the ratio a/A is made extremely small and thus is neglected. In this case the equation becomes:
$$h = s_2 h_2 - s_1 h_1$$ (v)

Sometimes the vertical tube of the micro-manometer is made inclined as shown in Figure (3-5):

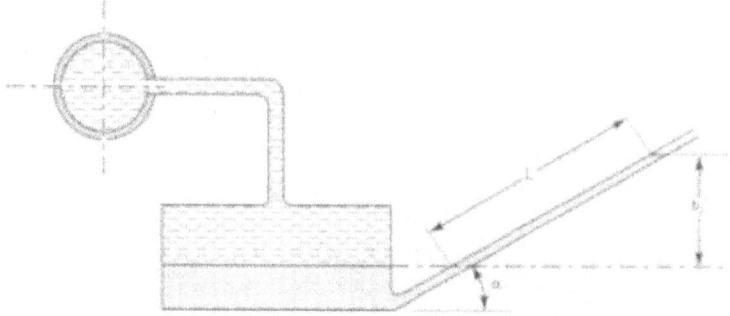

Fig (3-5) Inclined micro-manometer

This type of inclined micro-manometer is more sensitive than the vertical tube type. Due to inclination, the distance moved by the heavy liquid, in the narrow tube, will be comparatively more.

From the geometry of Figure (3-5), we find:
$$h_2/L = \sin\alpha$$
$$h_2 = L\,\sin\alpha$$

By substituting the value of h_2 in the micro-manometer equation, we can find out the required pressure in the pipe.

3.2.3 Differential manometer: It is a device for measuring the difference of pressures between two points in a pipe or in two different pipes.

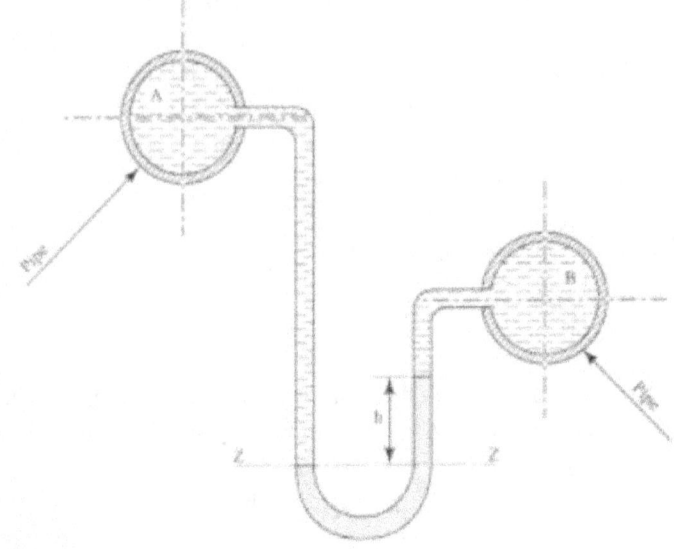

Fig (3-6) Differential manometer

A differential manometer consist of a U-tube, containing a heavy liquid, whose two ends are connected to the points, whose difference of pressure is required to be found out, as shown in Figure (3-6).

Now consider a differential manometer whose two ends are connected with two different points A and B at the same level.

Assuming the pressure at point A is more than at point B. The greater pressure at A will force the heavy liquid in the U-tube to move downwards. This downward movement of the heavy liquid, in the left side, will cause a corresponding rise to the heavy liquid in the right side.

Taking the horizontal surface Z-Z, at which the heavy liquid and light liquid meets, in the left side, as the datum line.

Let,
h = difference of the levels of the heavy liquid in the right and left sides (also known as the reading of the differential manometer) in mm,
s_1 = specific gravity of the light liquid in the pipes, and
s_2 = specific weight of the heavy liquid.

We know that the pressures in the left side and right side, above the datum line, are equal. Therefore, the difference of pressure in the two points A and B is:

$$h_A - h_B = h \times (\text{sp. gr. Of heavy liquid - sp. gr. Of liquid in pipes A and B})$$
$$= h \, (s_2 - s_1) \text{ head of water}$$

Sometimes the two pipes or the two points, whose difference of pressure is required to be found out are not at the same level; and at the same time the liquid flowing in the two pipes are different.

Now consider a differential manometer, whose two ends are connected to two different points A and B at different levels:

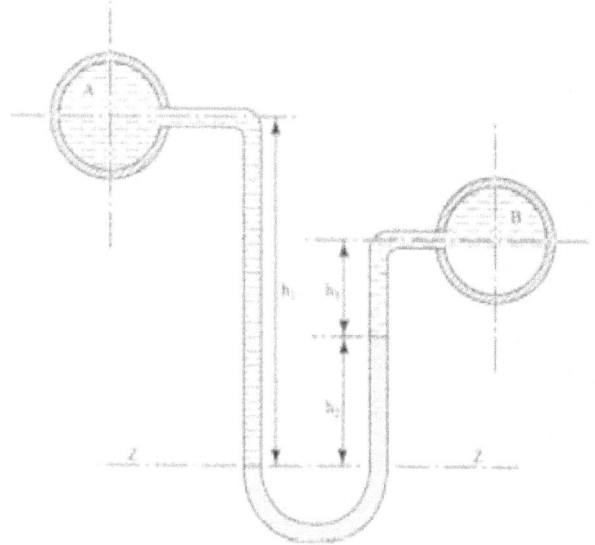

Fig (3-6) Differential manometer at different levels

Assuming the pressure at point A is more than that at point B; the greater pressure at A will force the heavy liquid to move downwards. Tis downward movement of the liquid, in the left side, will cause a corresponding rise of the heavy liquid in the right side as indicated in Figure (3-6).

The horizontal surface, at which the heavy and light liquid meet in the left side, is taken as a datum line. Let z-z be the datum line, in this case:

Let,

h_1 = height of light liquid in the left side above the datum line (mm),

h_2 = difference of levels of the heavy liquid in the right and left side (also known as reading of the differential manometer), (mm),

h_3 = height of the liquid in the right side above the datum line (mm),

h_A = pressure in the pipe A, expressed in terms of head of the liquid (mm),

h_B = pressure in the pipe B, expressed in terms of head of the liquid (mm),

s_2 = specific gravity of the heavy liquid, and

s_3 = specific gravity of the liquid in the right pipe (B).

As the pressures in the left side and right side, above the datum line, are equal, then

Pressure in the left side above the datum surface = $h_A + s_1 h_1$ (i), and

Pressure in the right side above the common surface = $s_2 h_2 + s_3 h_3 + h_B$ (ii)

$$h_A + s_1 h_1 = s_2 h_2 + s_3 h_3 + h_B$$

From the above equation, the values of h_A and h_B or their difference can be found.

3.2.4 Inverted differential manometer: This is a particular type of differential manometer, in which an inverted U-tube is used. An inverted differential manometer is used for measuring the difference of low pressures when accuracy is the prime consideration.

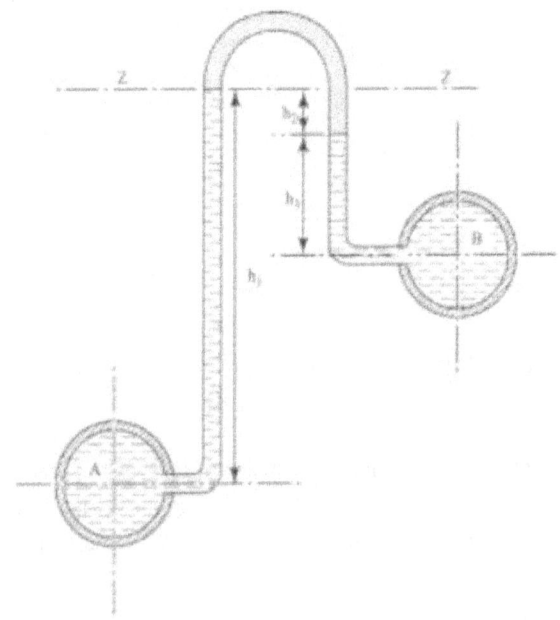

Fig (3-8) Inverted differential manometer

It consists of an inverted U-tube, containing a light liquid whose two ends are connected to the points whose difference of pressures is to be found out as shown in Figure (3-8).

Consider an inverted differential manometer, whose two ends are connected to two different points A and B. Let us assume that the pressure at point A is more than that at point B. The greater pressure at A will force the light liquid in the inverted U-tube to move upwards. This upward movement of liquid in the left side will cause a corresponding fall of the light liquid in the right side.

Taking z-z as the datum line:
Let,
h_1 = height of liquid in the left side above the datum line (mm),
h_2 = difference of levels of the light liquid in the right and left side (also known as reading of the differential manometer), (mm),
h_3 = height of the liquid in the right side below the datum line (mm),
h_A = pressure in the pipe A, expressed in terms of head of the liquid (mm),
h_B = pressure in the pipe B, expressed in terms of head of the liquid (mm),
s_1 = specific gravity of the liquid in the left side,
s_2 = specific gravity of the light liquid, and
s_3 = specific gravity of the liquid in the right side.

As the pressures in the left side and right side below the datum line are equal, hence:

Pressure in the left side below the datum line = $h_A - s_1 h_1$, and   (i)

Pressure in the right side below the datum line = $h_B - s_2 h_2 - s_3 h_3$ (ii)

Equating pressures in (i) and (ii), we get:

$$h_A - s_1 h_1 = h_B - s_2 h_2 - s_3 h_3, \text{ head of water}$$

From the above equation, the values of h_A and h_B or their difference can be found.

3.3 Mechanical Gauges

Whenever a very high fluid pressure is to be measured, a mechanical gauge is best suited for the purpose. A mechanical gauge is also used for the measurement of pressure in boilers or other pipes, where tube gauges cannot be conveniently used. There are many types of gauges available, but the principle on which they all work, is almost the same.

The following are two types of gauges that are important from the subject point of view:
1) Bourdon's tube pressure gauge, and
2) Diaphragm pressure gauge.

3.3.1 Bourdon's tube pressure gauge: Pressure above or below the atmospheric pressure may be easily measured, with the help of a Bourdon's tube pressure gauge.

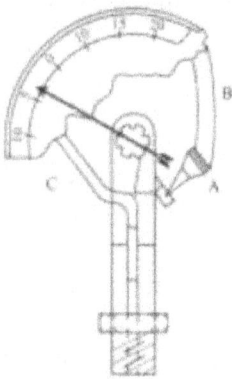

Fig (3-9) Bourdon's tube

Bourdon's tube pressure gauge, in its simplest form, consists of an elliptical tube ABC, bent into an arc of a circle as shown in Figure (3-9). This bent tube is called Bourdon's tube. When the gauge tube is connected to the fluid (whose pressure is to be found out) at C, the fluid under-pressure flows into the tube. The Bourdon's tube, as a result of the increased pressure, tends to straighten itself. Since the tube is encased in a circular cover, therefore it tends to become circular instead of becoming straight. With the help of simple pinion and sector arrangement, the elastic deformation of the Bourdon's tube rotates the pointer. This pointer moves over a calibrated scale, which directly gives the pressure.

3.3.2 Diaphragm pressure gauge: Pressure above or below the atmospheric pressure may also be found out easily with the help of a diaphragm in pressure gauge.

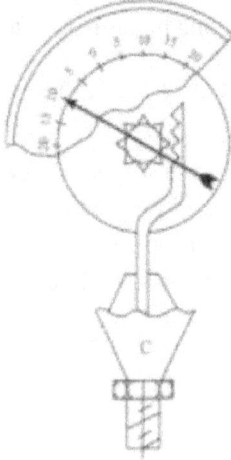

Fig (3-10) Diaphragm pressure gauge

A diaphragm pressure gauge, in its simplest form, consists of a corrugated diaphragm. When the gauge is connected to the fluid (whose pressure is to be found out) at C, the fluid under pressure causes some deformation of the diaphragm. With the help of some pinion arrangement the elastic deformation of the diaphragm rotates the pointer. This pointer moves over a calibrated scale, which directly gives the pressure. A diaphragm pressure gauge is generally used to measure relatively low pressures.

Check Your Knowledge
1. State the different principles of measurement of measurement of pressure.
2. Distinguish between piezometer and pressure gauge, when and where are they used.
3. Describe the different types of manometers.
4. State the difference between simple and differential manometers.
5. What are inclined manometers? For what purpose are these used in the laboratory?
6. What are mechanical gauges? In what circumstances are they used?
7. Write short notes on: (1) Bourdon's tube pressure gauge, (2) Diaphragm pressure gauge.

4. Fluids in Motion

4.1 Types of Flow in Pipes

When a fluid is flowing in a pipe, innumerable small particles get together and form a flowing stream. These particles, while moving, group themselves in a variety of ways, i.e. they may move in a regular formation just as disciplined soldiers do; or they may swirl and jostle like individuals in a disorderly mob.

The type of flow of a liquid depends upon the manner in which the particles unite and move.

4.1.1 Laminar flow: It is a flow in which the viscosity of the fluid is dominating over the inertia forces. It is more or less a theoretical flow, which rarely comes in contact with technical application, and is also known as viscous flow.

A laminar flow can be best understood by the hypothesis that the liquid moves in the form of concentric cylinders sliding one within the other as shown in Figure (4-1):

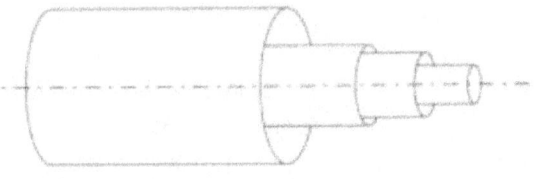

Fig (4-1) Laminar flow pipe

These concentric cylinders move like lamina. Such a flow, which takes place at very low velocities, is known as 'laminar flow'.

4.1.2 Turbulent flow: In this type of flow each liquid particle does not have a definite path and the paths of the individual particle cross each other.

It is a flow in which the inertia forces are dominating over the viscosity. It is a practical flow which comes in contact with technical applications.

In this flow the concentric cylinders diffuse or mix with each other and the flow is a disturbed one. Such a flow, taking place at high velocities, is known as 'turbulent flow'.

4.1.3 Uniform flow: In this type of flow the velocities of the liquid particles, at all sections of the pipe or channel, are equal.

4.1.4 Streamline flow: In this type of flow each liquid particle has a definite path and the paths of the individual particles do not cross each other.

4.1.5 Steady flow: In this type of flow the quantity of flowing liquid per second is constant.

4.2 Rate of Discharge

The quantity of a liquid, flowing per second through a section, of a pipe or a channel, is known as the rate of discharge or simply discharge. The discharge of a liquid is generally denoted by Q.

Considering a liquid flowing through a pipe:
Let,
a = cross sectional area of the pipe, and
υ = velocity of the liquid.
Then, the discharge is:
$$Q = \text{Area x Velocity} = a\ \upsilon \tag{i}$$

If the area is in square meters and velocity in m/s, then the discharge will have the unit:

$$Q = m^2 \text{ x m/s} = m^3/\text{sec}$$

$[1m^3 = 1000 \text{ litres}]$

4.3 Equation of Continuity of a Liquid Flow

If a liquid is continuously flowing through a pipe or a channel (whose cross-sectional area may or may not be constant) the quantity of the liquid passing per second is the same at all sections. This is known as the equation of continuity of a liquid flow:

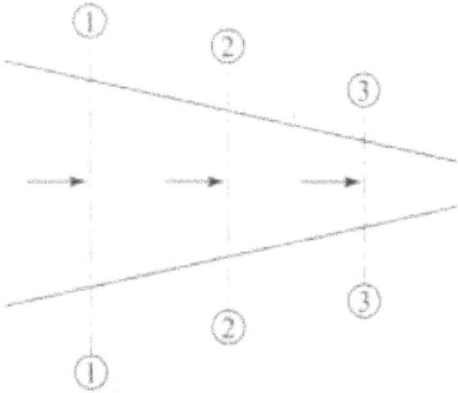

Fig (4-2) Liquid flowing through a pipe

Consider a tapering pipe, through which liquid is flowing as shown in Figure (4-2):
Let,
a_1 = area of the pipe at section 1-1, and
v_1 = velocity of the liquid at section 1-1.
Similarly,
a_2 = area of the pipe at section 2-2, and
v_2 = velocity of the liquid at section 2-2.
Similarly,
a_3 = area of the pipe at section 3-3,
v_3 = velocity of the liquid at section 3-3, and
w = specific weight of the liquid.

Weight of the liquid passing through section 1-1:

$$Q_1 = w\, a_1\, v_1$$

Similarly, weight of the liquid passing through section 2-2:

$$Q_2 = w\, a_2\, v_2$$

And weight of the liquid passing through the section 3-3:

$$Q_3 = w\, a_3\, v_3$$

Since the weight of liquid passing through the sections 1-1, 2-2 and 3-3 is the same,
Therefore,

$$Q_1 = Q_2 = Q_3$$

Or, $$w\, a_1\, v_1 = w\, a_2\, v_2 = w\, a_3\, v_3$$
Or, $$a_1\, v_1 = a_2\, v_2 = a_3\, v_3 = \ldots\ldots$$

This is known as the equation of continuity.

4.4 Energy of Liquid in Motion

Energy, in general, may be defined as the capacity to do work. Tough the energy may exist in many forms, yet the following are important from the subject point of view:

(1) Potential energy,
(2) Kinetic energy, and
(3) Pressure energy.

4.4.1 Potential energy of a liquid in motion: It is the energy, possessed by the liquid, by virtue of its position.

If a liquid particle is Z meters above the horizontal datum (arbitrarily chosen), the potential energy of the particle will be the Z metre-kilogram (briefly written as m-kg) per kg of liquid. Potential head of the liquid at that point will be the Z metres of the liquid.

4.4.2 Kinetic energy of a liquid in motion: It is the energy, possessed by the liquid particle, by virtue of its motion or velocity.

If a liquid particle is flowing with a mean velocity of v metres per second, then the kinetic energy of the particle will be $v^2/2g$ m-kg per kg of liquid. Kinetic head of the liquid, at that velocity, will be $v^2/2g$ meters of liquid.

4.4.3 Pressure energy of a liquid in motion: It is the energy, possessed by a liquid particle, by virtue of its existing pressure.

If a liquid particle is under a pressure 'p' kg per square metre, then the pressure energy of the particle will be p/w kg per kg of the liquid, where w is specific weight of the liquid. Pressure head of the liquid, under that pressure, will be p/w meters of the liquid.

4.4.4 Total energy of a liquid particle in motion: The total energy of a liquid, in motion, is the sum of its potential energy, kinetic energy and pressure energy.

Mathematically:

$$\text{Total energy. } E = Z + (v/2g) + (p/w) \text{ m-kg/kg of liquid}$$

4.4.5 Total head of a liquid particle in motion: The total head of a liquid particle in motion is the sum of its potential head, kinetic head and pressure head.

Mathematically,

$$\text{Total head, } H = Z + (v^2/2g) + (p/w) \quad \text{metres of liquid}$$

4.5 Bernoulli's Theorem

Bernoulli's theorem states: 'For a perfect incompressible liquid, flowing in a continuous stream, the total energy of a particle remains the same, while the particle moves from one point to another'. This statement is based on the assumption that there are no losses due to friction in the pipe.

Mathematically,

$Z + v^2/2g + p/w = \text{constant}$

where,

Z = potential energy,

$v^2/2g$ = kinetic energy, and

p/w = pressure energy

4.5.1 Proof of Bernoulli's theorem: Consider a perfect incompressible liquid, flowing through a non-uniform pipe as shown in Figure (4-3):

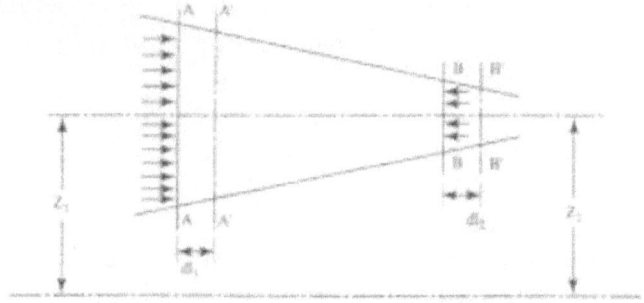

Fig (4-3) Flow through non-uniform pipe

Taking two sections AA and BB of the pipe, and assuming that the pipe is running full and there is a continuity of flow between the two sections.

Let,

Z_1 = height of AA above the datum,

p_1 = pressure of AA,

v_1 = velocity of liquid at AA, and

a_1 = area of pipe at AA.

Similarly,

Z_2 = height of BB above the datum,

p_2 = pressure of BB,

v_2 = velocity of liquid at BB, and

a_2 = area of pipe at BB.

Let the liquid between the two sections AA and BB move to A$^|$ A$^|$ and B$^|$ B$^|$ through very small lengths dL_1 and dL_2 as shown in Figure (4-3). This movement of the liquid between AA and BB is equivalent to the movement of the liquid between AA and A$^|$A$^|$ to BB and B$^|$B$^|$. The remaining liquid between A$^|$A$^|$ and BB is unaffected.

Let W be the weight of the liquid between AA and $A^|A^|$. Since the flow is continuous, therefore:

$$W = w \, a_1 \, dL_1$$

Or,

$$a_1 \, dL_1 = W /w \qquad\qquad (i)$$

Similarly,

$$a_2 \, dL_2 = W /w$$

Therefore,

$$a_1 \, dL_1 = a_2 \, dL_2 \qquad\qquad (ii)$$

Now work done by pressure at AA, in moving the liquid to $B^|B^| = - p_2 \, a_2 \, dL_2$ (*minus sign is taken as the direction of p_2 is opposite to that of p_1*).

Therefore,

Total work done by the pressure $= p_1 \, a_1 \, dL_1 - p_2 \, a_2 \, dL_2$

$\qquad\qquad\qquad\qquad = p_1 \, a_1 \, dL_1 - p_2 \, a_2 \, dL_1 \qquad$ (as $a_1 \, dL_1 = a_2 \, dL_2$)

$\qquad\qquad\qquad\qquad = a_1 \, dL_1 \, (p_1 - p_2)$

$\qquad\qquad\qquad\qquad = W/w \, (p_1 - p_2) \qquad$ (as $a_1 \, dL_1 = W/w$)

Loss of potential energy $= W \, (Z_1 - Z_2)$

And gain in kinetic energy $= W \, (v_2^2/2g - v_1^2/2g) = W/2g \, (v_2^2 - v_1^2)$

We know that:

$\qquad\qquad$ Loss of potential energy + work done by pressure = Gain in kinetic energy

Or,

$$W \, (Z_1 - Z_2) + W/w \, (p_1 - p_2) = W/2g \, (v_2^2 - v_1^2)$$

Or,

$$Z_1 - Z_2 + p_1/w - p_2/w = v_2^2/2g - v_1^2/2g$$

Or,

$$Z_1 + v_1^2/2g + p_1/w = Z_2 + v_2^2/2g + p_2/w$$

which proves Bernoulli's theorem.

4.5.2 Limitation of Bernoulli's theorem: The Bernoulli's theorem or equation has been derived on some assumptions, which are rarely possible. Thus the Bernoulli's theorem has the following limitations:

(1) The Bernoulli's equation has been derived under the assumption that the velocity of every liquid particle, across any cross section of a pipe, is uniform. But in actual practice, it is not so. The velocity of liquid particle in the centre of the pipe is at its maximum and gradually

decreases towards the walls of the pipe due to the pipe friction. Thus while using the Bernoulli's equation, only the mean velocity of the liquid should be taken into account.

(2) The Bernoulli's equation has been derived under the assumption that no external force, except the gravity force, is acting on the liquid. But in actual practice it is not so. There are always some external forces (such as pipe friction etc...) acting on the liquid, which affects the flow of the liquid. Thus, on using Bernoulli's equation all such external forces should be neglected. But if some energy is supplied to, or extracted from, the flow, then this energy should be taken into account.

(3) The Bernoulli's equation has been derived under the assumption that there is no loss of energy of the liquid particle while flowing. But in actual practice it is rarely so. In a turbulent flow, some kinetic energy is converted into heat energy and in a viscous flow some energy is lost due to shear forces. Thus, while using Bernoulli's equation all such losses should be neglected.

(4) If the liquid is flowing in a curved path, the energy due to centrifugal force should be taken into account.

4.6 Practical Application of Bernoulli's Equation

The Bernoulli's theorem or equation is the basic equation, which has the widest application in hydraulics and applied hydraulics. Since this equation is applied for the derivation of many formulae in hydraulics its clear understanding is very essential.

Bernoulli's theorem is applied to the following hydraulic devices:
(a) Venturimeter,
(b) Orificemeter, and
(c) Pitot tube.

4.6.1 Venturimeter:

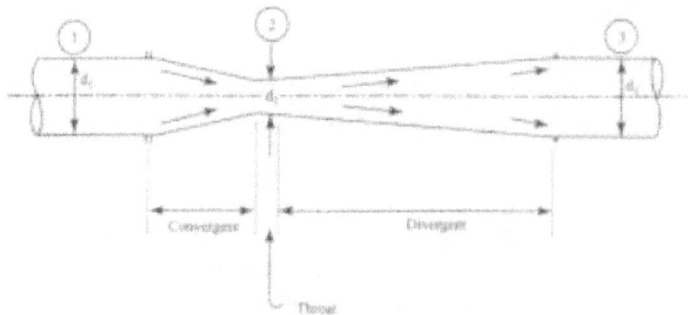

Fig (4-4) Venturimeter

The venturimeter is an apparatus for finding out the discharge of a liquid flowing in a pipe. It is one of the practice applications of Bernoulli's equation.

A venturimeter in its simplest form consists of the following three parts:
(i) Convergent cone,
(ii) Throat, and
(iii) Divergent cone

(i) **Convergent cone:** It is a short pipe, which converges from a diameter d_1 (diameter of the pipe in which the venturimeter is fitted) to a small diameterd_2.

The convergent cone is also known as the inlet of the venturimeter. The slope of the converging sides is between 1 in 4 or 1 in 5 as shown in Figure (4-4).

(ii) **Throat:** It is a small portion of circular pipe, in which in the diameter d_2 is kept constant.

(iii) **Divergent cone:** It is a pipe, which diverges from a diameter d_2 to a larger diameterd_1. The Divergent cone is also known as the outlet of the venturimeter. The length of the divergent cone is about 3 to 4 times than that of the convergent cone.

The liquid, while flowing through the venturimeter, is accelerated between the sections (1) and (2) (i.e. flowing through the convergent cone). As a result of this acceleration the velocity of liquid at section 2 (i.e. at the throat) becomes higher than that at section(1).

This increase in velocity results in decreasing the pressure at section (2) considerably. If the pressure head at the throat falls below the separation head (which is 2.5 meters of water) then there will be a tendency of separation of the liquid flow.

In order to avoid the tendency of the separation, at throat, there is always a fixed ratio of the diameter of the throat and the pipe (i.e. d_2/d_1). This ratio varies from ¼ to ¾ but the most suitable value is 1/3 to ½.

The liquid, while flowing through the venturimeter, is accelerated (i.e. retarded) between the sections (2) and (3) (i.e. while flowing through the divergent cone). As a result of this retardation the velocity of liquid decreases which, consequently, increases the pressure.

If the pressure is rapidly recovered, then there is every possibility for the stream of liquid to break away from the walls of the meter due to boundary layer effects.

In order to avoid the tendency of breaking away the streams of liquid, the divergent cone is made sufficiently longer.

Another reason for making the divergent cone longer, is to minimise the frictional losses.

For these reasons, the divergent is 3 to 4 times longer than the convergent cone.

4.6.2 Discharge through a venturimeter:

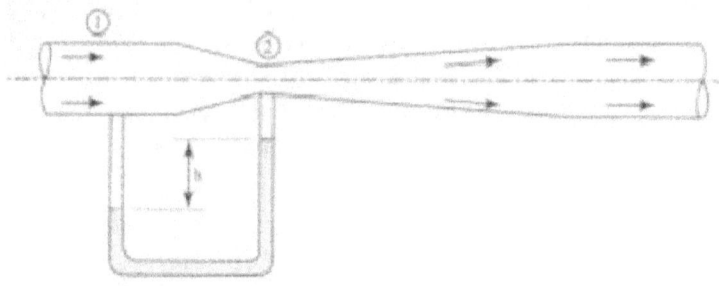

Fig (4-5) Liquid flowing through a venturimeter

Consider a venturimeter, through which some liquid is flowing as shown in Figure (4-5):
Let,
p_1 = pressure at section 1,
v_1 = velocity of water at section 1,
Z_1 = datum head at section 1, and
a_1 = area of the venturimeter at section 1.
Similarly,
p_2 = pressure at section 2,
v_2 = velocity of water at section 2,
Z_2 = datum head at section 2, and
a_2 = area of the venturimeter at section 2.

Applying Bernoulli's equation at sections 1 and 2:

$$Z_1 + v_1^2/2g + p_1/w = Z_2 + v_2^2/2g + p_2/w \qquad (i)$$

Let the datum line passes through the axis of the venturimeter as shown in Figure (4-5),
As,

$$Z_1 = 0 \quad \text{and} \quad Z_2 = 0$$

$$v_1^2/2g + p_1/w = v_2^2/2g + p_2/w$$

Or, $p_1/w - p_2/w = v_2^2/2g - v_1^2/2g$ (ii)

Since the discharge at sections 1 and 2 is continuous:

$$a_1\, v_1 = a_2\, v_2$$

$$v_1 = (a_2\, v_2) / a_1$$

$$v_1^2 = (a_2^2\, v_2^2) / a_1^2 \qquad\qquad\text{(iii)}$$

Substituting the above value of v_1^2 in equation (ii):

$$p_1/w - p_2/w \;= v_2^2/2g - (a_2^2 / a_1^2) \times v_2^2/2g$$

$$= v_2^2/2g \,[1 - (a_2^2 / a_1^2)]$$

$$= v_2^2/2g \,[(a_1^2 - a_2^2) / a_1^2]$$

Since $(p_1/w - p_2/w)$ is the difference between the pressure heads at sections 1 and 2 and if the pipe is horizontal, then this difference represents the venture head and is denoted by 'h':

Or, $h = v_2^2/2g \,[(a_1^2 - a_2^2) / a_1^2]$

Or, $v_2^2 = 2g \,[a_1^2 / (a_1^2 - a_2^2)]$

Therefore, $v_2^2 = \sqrt{2gh} \;[a_1 / \sqrt{(a_1^2 - a_2^2)}\,]$

And since the discharge through a venturimeter, Q = Coefficient of venturimeter (coefficient of discharge) x $a_2\, v_2$

$$= C\, a_2\, v_2 = [(C\, a_1\, a_2) / \sqrt{(a_1^2 - a_2^2)}\,]\,(\sqrt{2gh}\,)$$

Note: The venture head h, in the above equation, has been taken in terms of the liquid head. But in actual practice, this head is given as mercury head. In each case the mercury head should be converted into the liquid head.

If the given liquid is water, the mercury head can be converted into water head by multiplying the mercury head by the specific gravity of mercury, minus the specific gravity of water (i.e. 13.6 - 1 = 12.6).

Sometimes oil is being discharged through the venturimeter. In such a case the venture head should be taken in terms of oil head, i.e.

$$h = (13.6 - w) / w \text{ x head of mercury}$$

where,

h   = venture head in terms of liquid head,

13.6 = specific gravity of mercury, and

w   = specific weight of the oil.

4.6.3 Inclined venturimeter: Sometimes a venturimeter is fitted to an inclined (or even a vertical) pipe as shown in Figure (4 - 6):

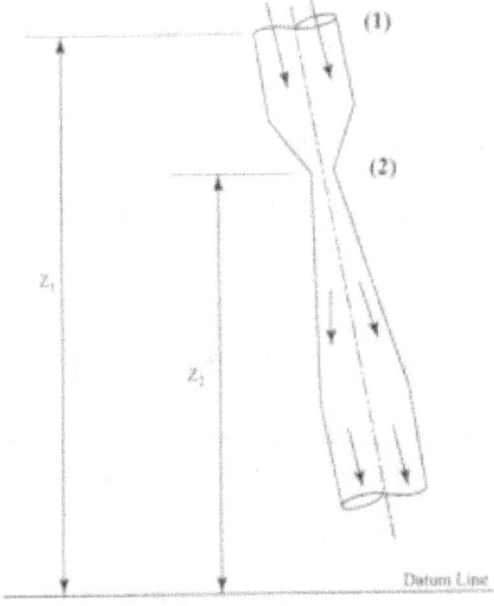

Fig (4-6) Inclined venturimeter

The same formula $\{Q = [(C\ a_1\ a_2)/\sqrt{(a_1^2 - a_2^2)}\ ]\ (\sqrt{2gh}\ )\}$ for discharge, through the venturimeter holds good. The discharge through an inclined venturimeter Z_1 may also be found out, first by finding out the velocity at either sections (by using Bernoulli's equation) and then by multiplying the velocity with the respective area of flow.

4.7 Forced Vortex Flow

It is a type of vortex flow, in which the vessel, containing a liquid, is forced to rotate about a fixed vertical axis with the help of a torque. If the applied torque is removed, the rotational motion will be slowly destroyed due to viscous and turbulent shear forces in the liquid.

Consider a cylindrical vessel containing initially a liquid up to AA as shown in Figure (4 - 7):

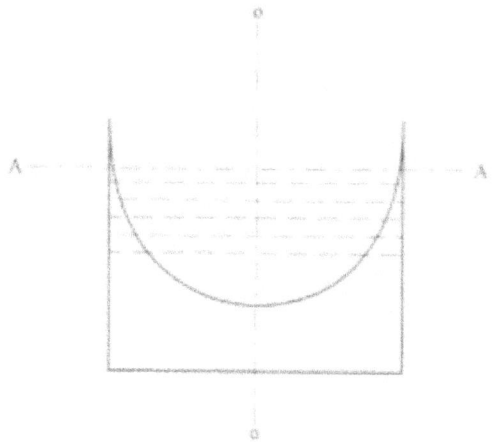

Fig (4-7) Vortex flow

Let the vessel be rotated about its vertical axis o-o. The liquid surface in the vessel is depressed down at the axis of rotation, and rises up near the wall of the vessel on all sides.

If the vessel is revolved with an increased angular velocity, the liquid will depress down to a greater extent at its axis of rotation, and rises up to a greater extent near the wall of the vessel.

By increasing the velocity of rotation, the liquid inside the vessel will start to spill out. By further increasing the velocity of rotation, the depression will increase until the liquid's axial depth reaches the zero value.

Consider the case when a vessel containing a liquid is revolving about its axis to form a curved surface:
Let,
N = rotation of vessel in r.p.m., and
ω = angular velocity of the vessel in radians / sec.

Now consider a particle P on the liquid surface of any radius χ from the axis of rotation, as shown in Figure (4 - 8):

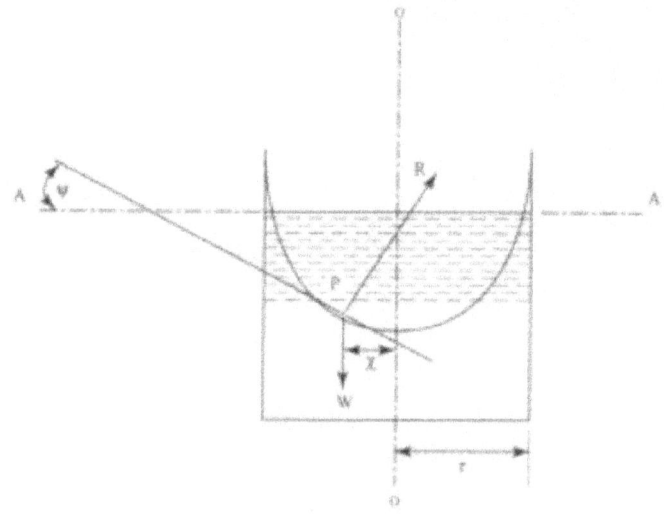

Fig (4-8) Vessel containing liquid revolving about its vertical axis

The particle P is in equilibrium under the action of the following forces:

(1) height of the particle W,
(2) centrifugal force, and
(3) reaction R exerted by the liquid on the particle.

Since, the centrifugal force = W/g $(\omega^2 \chi)$ (i)

This force will act horizontally outwards and the weight will act vertically downwards.

Let the tangent at P make an angle ψ with the horizontal as shown in Figure (4-8).

Resolving the weight of the particle and the centrifugal force along tangent at P and equating the same to get:
$$W/g \, (\omega^2 \chi) \cos \psi = W \sin \psi$$
Or,
$$\tan \psi = W/g \, (\omega^2 \chi) / W = \omega^2 \chi /g$$
Or,
$$dy/d \chi = \omega^2 \chi /g \qquad \text{(as } \tan \psi = dy/d \chi)$$

Integrating the above equation,
$$y = (\chi^2 \omega^2 /2g) + C$$
Where C is the constant of integration.

If the lowest point of the curved liquid surface is chosen as an origin then by putting $\chi = 0$ and $y = 0$, then C=0.
Therefore,

$$y = \omega^2 \chi^2 / 2g$$

This is the equation of a parabola, which means that the surface of the liquid is parabolic.

If the liquid particle taken at radius 'r' from the axis of rotation, then the above equation becomes:

$$y = \omega^2 r^2 / 2g$$

4.7.1 Free vortex: In this type of flow the liquid particles describe circular paths about a fixed vertical axis, without any external force acting on the particles. The common example of a free vortex occurs when the water escapes, through an orifice, at the bottom of a wash basin.

Consider a liquid having a free vortex as shown in Figure (4 - 9):

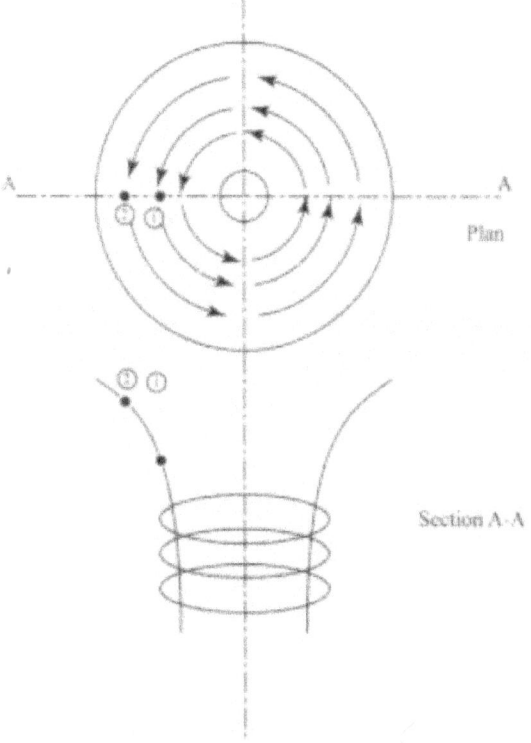

Fig (4-9) Free vortex

Let the liquid move with an angular velocity of ω radians/sec. Consider two liquid particles 1 and 2 on the liquid surface as shown on Figure (4 - 9):

Let,

r_1 = radius of particle 1,

r_2 = radius of particle 2,

υ_1 = tangential velocity of particle 1,

υ_2 = tangential velocity of particle 2, and

ω = angular velocity of both particles.

Tangential velocity of particle 1,
$$\upsilon_1 = \omega \, r_1 \quad \text{or,} \quad \omega = \upsilon_1/r_1 \qquad \text{(i)}$$
Similarly,
$$\upsilon_2 = \omega \, r_2 \quad \text{or,} \quad \omega = \upsilon_2/r_2 \qquad \text{(ii)}$$

Since both particles have the same angular velocity ω radians/sec, therefore by equating (i) and (ii) we get:
$$\upsilon_1 / r_1 = \upsilon_2 / r_2$$

Applying Bernoulli's equation to points 1 and 2:
$$\upsilon_1^2/2g + p_1/\omega = \upsilon_2^2/2g + p_2/\omega \quad \text{(taking } Z_1 = Z_2)$$
Or,
$$p_1/\omega - p_2/\omega = \upsilon_2^2/2g - \upsilon_1^2/2g$$
$$(p_1 - p_2) / \omega = (\upsilon_2^2 - \upsilon_1^2) / 2g$$
$$(p_2 - p_1) / \omega = (\upsilon_1^2 - \upsilon_2^2) / 2g \qquad \text{(taking minus outside)}$$

4.7.2 Modified 'Free Vortex' equation (application): Replacing the blade profile in a viscous flow by a series of revolving cylinders having a suitable rotational velocity, and arranged along the centre-line of the blade. Each of these cylinders will have its own vortex field created around it.

Taking the cylinder diameter as a representative of the mean thickness of the blades, and the vorticity as the circulation for a given length of the profile contour.

Free vortex is considered as an idealisation of a thin cylinder of fluid, which in limit cannot sustain forces. The condition that a free vortex does not sustain forces is used to determine its motion.

Consider a viscous flow, the vortex filaments in an incompressible fluid is proposed to satisfy the following rotational velocity relation:

$$\upsilon_n \cdot r_n = K$$

Where,
v_n = circumferential velocity,
r_n = vortex radius, and
K = constant.

The value m = 1 refers to the ideal case of frictionless flow. Although the exponent (m) changes as a function of the radius (r), it carries greater values with increased distances ahead a set of blades, while the velocity at positions closer to the set of blades will possess a decreasing value due to friction losses.

The exponent (m) changes as a function of the radius (r_n), especially in ranges of large distances ahead set of blades. By practice, the approximate value of (m) for air at normal temperature was found to be: m = 1.15.

4.8 Similitude of Fluid Flow

Predicting the performance of structures or machines is done by preparing models and testing them in a laboratory; so as to form some opinion about the working and behaviour of the proposed hydraulic structures, after their completion or actual installation. The structure, of which the model is prepared, is known as a prototype and the model is known as a scale model or simple model.

Dimensional analysis applied to problems of similitude in hydraulics proved to be a useful tool in many instances. It discloses functional relationship among quantities involved and establishes dimensionless criteria of flow.

The method of procedure is proposed by Buckingham (π Theorem), which states: the principle of dimensional analysis requires that all the terms of a correct and complete physical equation shall have the same dimensions. This implies that the object to be studied by dimensional analysis should be known well enough to permit assumption of the physical quantities expected to affect the phenomenon under consideration.

4.8.1 Kinematic similarity: The kinematic similarity is said to exist between the model and prototype if both have identical motion or velocity. Kinematic similarity can also exist between the model and prototype, if the ratio of the corresponding velocities at corresponding points is equal:
Let,
V_1 = velocity of liquid in the prototype at point 1,
V_2 = velocity of fluid in the prototype at point 2,
v_1 = velocity of fluid in the model at point 1,

υ_2 = velocity of fluid in the model at point 2, and

V_r = velocity ratio of the prototype and the model.

If kinematic similarity exists between prototype and model we have:

$$V_r = V_1/\upsilon_1 = V_2/\upsilon_2 = V_3/\upsilon_3 = \ldots..$$

4.8.2 Dynamic similarity: The dynamic similarity is said to exist between the model and prototype if both have identical forces. Or in other words, the dynamic similarity is said to exist between the model and prototype if the ratio of the corresponding forces acting at corresponding points is equal:

Let,

F_1 = force acting in the prototype at point 1,

F_2 = force acting in the prototype at point 2,

f_1 = corresponding force acting on the model at point 1,

f_2 = corresponding force acting on the model at point 2, and

F_r = force ratio of the prototype and the model.

If the dynamic similarity exists between prototype and true model we have:

$$F_r = F_1/f_1 = F_2/f_2 = F_3/f_3 = \ldots..$$

4.9 Reynolds Number

The Reynolds number is the ratio of inertia force and viscous force. If viscous force is the predominating force in the model and prototype, dynamic similarity is said to exist between the two resulting in equal Reynolds number for model and prototype:

Reynolds number, Re = Inertia force / Viscous force

= mass x acceleration / shear stress x cross section area

= [volume x mass density x (velocity/time)] / shear stress x cross sectional area

= [(velocity/time) x mass density x velocity] / shear stress x cross sectional area

= [(cross sectional area x velocity) x mass density x velocity] / shear stress x cross section area

= $(V \rho V) / \mu (d\upsilon/dy) = (V \rho V) / (\mu V/L)$

= $(V\rho) / (\mu/L)$

= $V / (\upsilon/L)$ (since, $\mu / V^2 = \upsilon$)

4.10 Froude's Number

Froude's number is the ratio of inertia force and gravity force. If gravity force is the predominating force in a model and prototype, dynamic similarity is said to exist between the two, and Froude's number for the model and prototype is the same:

Ratio of forces = Inertia force / Gravity force
= mass x acceleration / mass x acceleration due to gravity
= (V/T) /g = V / (Tg)
= V / (L/v) g = V^2/ Lg

The Froude's number is the square root of this ratio:

$$F_n = \sqrt{(V^2/ Lg)} = V/\sqrt{Lg}$$

4.11 Mach's Number or Cauchy's Number

The Mach's or Cauchy number is the ration of inertia force and the elastic force. If the elastic force is predominating force in the model and the prototype, dynamic similarity is said to exist between the two, and the Mach's number for the model and the prototype is the same:

Ratio of these two forces = Inertia force / Elastic force
= mass x acceleration/elastic stress x area
= [(volume x mass density) x (velocity/time)] / elastic stress x area
= [(cross sectional area x velocity) x mass density x velocity] / elastic stress x area
= (velocity x mass density x velocity) / elastic stress x area
= (V ρ V) / K = V$^2\rho$ / K

Mach's number is the square root of this ratio:

$$Ma = \sqrt{(V^2\rho/K)} = \sqrt{(V^2)} / \sqrt{(K/\rho)} = V/ \sqrt{(K/\rho)}$$

4.12 Weber Number

Weber's number is the ratio of inertia force to the surface tension force. If the surface tension force is the predominating force in the model and prototype, dynamic similarity is said to exist between the two, and the Weber's number for the model and prototype is the same:

Ratio of these two forces = Inertia force / surface tension force
= (mass x acceleration) / (surface tension per unit length x length)
= [(volume x mass density) (velocity / time)] / [(surface tension per unit length x length)]
= [(volume/time) x mass density x velocity] / [surface tension per unit length x length)]
= [(cross-sectional area x velocity) x mass density x velocity] / [surface tension per unit length x length]
= $(L^2 \, V \, \rho \, V) / (\sigma \, L) = (\rho \, L \, V^2) / \sigma$

Weber's number is the square root of this ratio:

$$W_N = \sqrt{(\rho \, L \, V^2) / \sigma} = \sqrt{(V^2)/(\sigma/\rho L)} = V / \sqrt{\sigma/\rho L}$$

Check Your Knowledge

1. What do you understand by the term vortex flow? Name the various types of vortex flow.
2. Distinguish between forced and free vortex.
3. Differentiate between a free vortex and forced vortex.
4. State and prove Bernoulli's theorem for flow of liquids.
5. Derive Bernoulli's equation from first principles for study of incompressible fluids.
6. Derive an expression for discharge through a venturimeter with a neat sketch.
7. Sketch a venturimeter and state why a certain angle of divergence is to be maintained.

Exercise

1. A model of airship of scale 1: 100 was tested in water at 5 meters/sec. Find the velocity of the actual airship, if the kinematic viscosity of air is 13 times that of water. [**Ans.** 38.46 m/sec]

2. A channel model 25 cm deep is discharging water flowing with a velocity of 1.5 meters/sec. Find the velocity of water in the channel 4 meters deep, if the model has dynamic similarity with its prototype and the flow is governed by Froude's law. [**Ans.** 6 m/sec]

3. An aerofoil moves at 650 kilometres per hour through still air at 20°C. If the elastic stress and density of air at this temperature is 21 kg/cm^2 and 0.126 kg/m^3, find the Mach's number. [**Ans.** Ma = 0.14]

4. A1:60 model of an aeroplane is tested in water at 10 metres/sec. If the kinematic viscosity of air and water is 0.15 stoke and 0.01 stoke respectively, find the velocity of the actual aeroplane. [**Ans.** 40 m/sec]

5. In exercise (4), if the pressure drop of the model during test is 2.0 kg/cm^2, find the corresponding pressure drop in the prototype. The water is 800 times denser than air. [**Ans.** 0.01 kg/cm^2]

5. Flow Measurements

5.1 Average Velocity of Flow

A number of methods are employed to ascertain the average velocity of flow, but the following are important from the subject point of view:

(1) Pitot tube, and
(2) Current meter.

5.1.1 Pitot tube: A pitot tube is an instrument used to determine the velocity of flow at a required point in a pipe or stream:

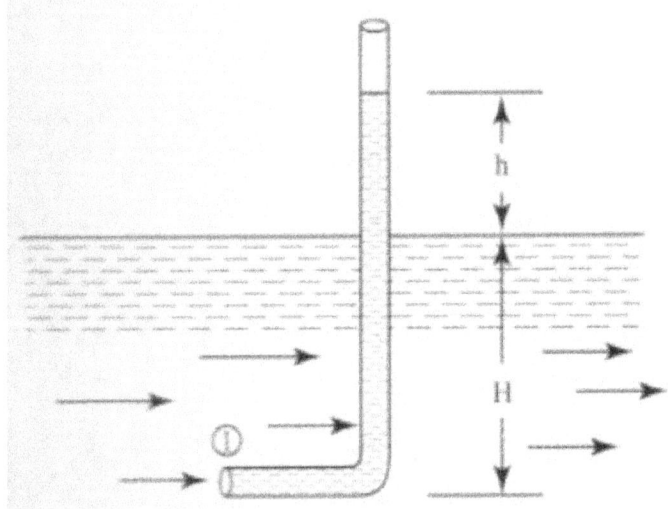

Fig (5-1) Pitot tube

In its simplest form a Pitot tube consists of a glass tube bent through 90° as shown in Figure (5-1). The lower end of the tube faces the direction of the flow; the liquid rises up in the tube due to the pressure exerted by the flowing liquid. By measuring the rise of liquid in the tube, we can find out the velocity of the liquid flow.

Let,

h = height of liquid in the Pitot tube above the surface,

H = depth of the tube in the liquid, and

υ = velocity of the liquid.

Applying Bernoulli's equation for the sections 1 and 2,

$$H + h = H + \upsilon_2^2/2g \qquad \text{(as } Z_1 = Z_2)$$

$$h = v_2^2/2g$$

Or,

$$v = \sqrt{2gh}$$

It has been experimentally found that if the Pitot tube is placed, with its nose facing side way, in the flow, there will be no rise of the liquid in the tube. But if a Pitot tube is placed, with its nose facing down, the liquid level in the tube will be depressed by an amount equal to h, such that,

$$h = \sqrt{v_2^2/2g}$$

Where,

v = velocity of the liquid flow

5.1.2 Current meter: It is an instrument to determine the velocity of flow at a required point in a flowing stream. Though there are many types of current meters available in the market, their basic principle is the same.

A current meter consists of a wheel containing blades or cups, which are rotated by the flowing water. The number of rotation of the wheel, within a certain time, depends upon the velocity of water. An electric current is passed from the battery to the wheel by means of a wire. A rotation of the wheel makes and breaks the electric circuit. Thus, by counting the breaks with time elapsed, the velocity of the flowing water is obtained.

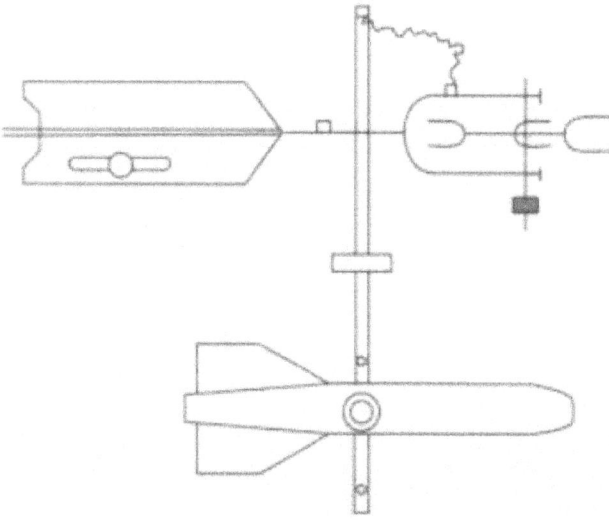

Fig (5-2) Current meter

The meter is suspended by means of a fine cable and lowered to the required depth. The current meter is free to move about its horizontal and vertical axis, so that it can adjust itself with the direction of the water flow.

5.1.3 Rating of current meters: The process of obtaining the relationship between the number of revolutions of the cups of the current meter in unit time and the velocity of water flowing past the meter is known as the calibration of the current meter or rating of the current meter.

The current meter rating station consists of a long straight channel fitted with water at rest. A pair of carefully levelled rails is laid one on each bank of the channel. These rails form a track for an electrically driven car. The current meter is hung from the car, which is driven along the track. In this way the current meter is towed through the water at various selected speeds and number of signals is recorded at the various speeds. A curve is then drawn with the help of known velocity and the signals recorded. The curve so obtained is known as a rating curve. Sometimes instead of plotting a rating curve, a rating table is prepared, which directly gives the velocity of water with the corresponding signals.

5.1.4 Precautions for rating the current meter: The following precautions are necessary for rating the current meters:

(1) The water in the rating tank must be allowed to come completely to rest after each run and before the beginning of the next.

(2) The current meter must not be allowed to approach too closely to the sides or bottom of the rating channel.

(3) The rating channel should not be less than 2 meter wide and 1.5 meter deep. It should be long enough to permit a clear run at a constant speed for at least 15 meters.

(4) The current meter should be supported by the same type of rod or cable, which is intended to be used during the velocity observation in the field.

5.2 Orifices

An opening, in a vessel, through which the liquid flows out, is known as an orifice. This hole or opening is called an orifice, so long as the level of the liquid on the upstream side is above the top of the orifice. The usual purpose of an orifice is the measurement of flow.

An orifice may be in a vertical side of the vessel or in the horizontal base. But the former is more common.

5.2.1 Types of orifices: There are many types of orifices, depending upon their size, shape or nature of discharge. But the following are important from the subject point of view:

(1) According to size: (i) Small orifice, and (ii) Large orifice.

(2) According to shape: (i) Circular orifice, (ii) Rectangular orifice, and (iii) Triangular orifice.

(3) According to shape of the edge: (i) Sharp-edged orifice, and (ii) Bell-mounted orifice.

(4) According to nature of discharge: (i) Fully submerged orifice, and (ii) Partly submerged orifice.

Before introducing the details of flow through all types of orifices, the following physical definitions are presented:

Jet of water: The stream of a liquid, which flows out of an orifice, is known as the 'jet of water'.

Vena Contracta: Consider a tank fitted with an orifice as shown in Figure (5-3)

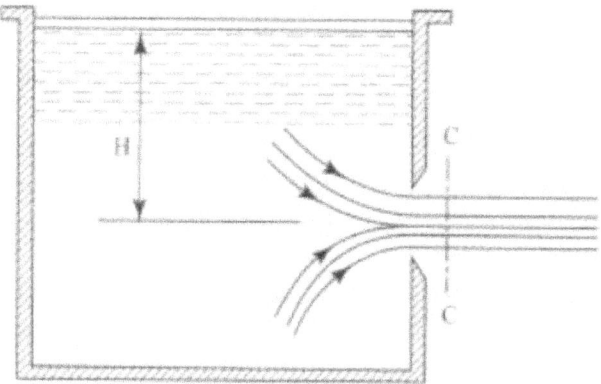

Fig (5-3) Tank with an orifice

The liquid particles will move towards the orifice from all directions.

Few of the particles move downward and take a turn to enter into the orifice and finally flow through it. The liquid particles will lose some energy, while taking the turn, to enter into the orifice. It was observed that the jet leaving the orifice gets contracted. The maximum contraction takes place at a section slightly on the downstream side of the orifice, where the jet is more or less horizontal. Such a section is known as 'Vena Contracta'.

5.3 Hydraulic Coefficients

The following three coefficients are known as hydraulic coefficient or orifice coefficients:

(1) Coefficient of concentration,
(2) Coefficient of velocity, and
(3) Coefficient of discharge.

5.3.1 Coefficient of contraction: The ratio of area of the jet, at vena contracta, to the area of the orifice is known as coefficient of contraction. Mathematically, the coefficient of contraction is:

$$C_c = \text{Area of jet at Vena Contracta / Area of Orifice}$$

The value of coefficient of contraction varies slightly with the available head of the liquid, size and shape of the orifice. An average value of C_c is about 0.64

5.3.2 Coefficient of velocity: The ratio of actual velocity of the jet, at vena contracta, to the theoretical velocity is known as coefficient of velocity. Mathematically, the coefficient of velocity is:

$$C_v = \text{Actual velocity of vena contracta / Theoretical velocity}$$

The difference between the two velocities is due to friction of orifice. The value of coefficient of velocity varies very slightly with different shapes of edges of an orifice (i.e. very small for sharp-edged orifices). An average value of C_v is about 0.97

5.3.3 Coefficient of discharge: The ratio of actual discharge through an orifice to the theoretical discharge is known as coefficient of discharge. Mathematically, the coefficient of discharge is:

$$C_d = \text{Actual discharge / Theoretical discharge} =$$
$$= \text{(Actual velocity x Actual area) / (Theoretical velocity x Theoretical area)}$$
$$= \text{(Actual velocity / Theoretical velocity) x (Actual area / Theoretical area)}$$
$$= C_v \times C_c$$

The value of coefficient of discharge varies with the values of C_c and C_v. An average value for coefficient of discharge is about 0.62

5.3.4 Experimental determination of hydraulic coefficients:

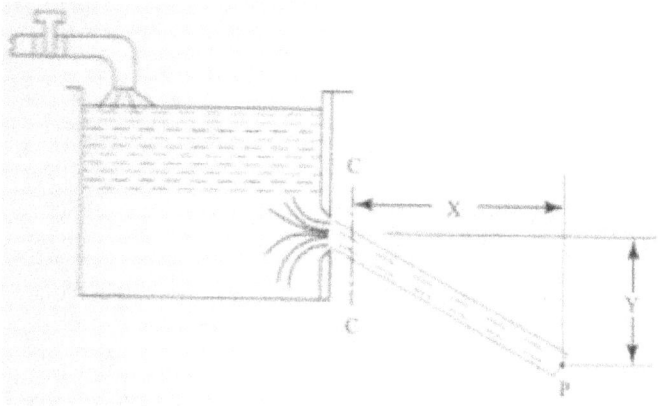

Fig (5-4) Tank with water at constant level

Consider a tank containing water at constant level, maintained by a constant level, maintained by a constant supply, as shown in Figure (5- 4):

Let the water flow out of the tank through an orifice, fitted in one side of the tank. Examining a particle of water in the jet at point P and assuming section C-C to represent the point of vena contract:

Let,

H = constant water head,

χ = horizontal distance between C-C and P,

y = vertical distance between C-C and P,

υ = velocity of the jet, and

t = time taken, in seconds, by the particle to reach from C-C to P.

Since,

$$y = \tfrac{1}{2}\, g\, t^2 \qquad\qquad (i)$$

(where, g is the gravitational acceleration), and

$$\chi = \upsilon \times t$$

Therefore,
$$t = \chi / \upsilon \qquad\qquad (ii)$$

Substituting the values of t in equation (i) we get:

$$y = \tfrac{1}{2}\, g\, (\chi / \upsilon) \qquad\qquad (ii)$$

$$2y / g = (\chi / \upsilon)^2$$

$$\upsilon = \sqrt{g\chi^2 / 2y} \qquad\qquad (iii)$$

This is an equation with a parabolic shape.

As the theoretical velocity of a particle is:

$$v_{th} = \sqrt{2gH}$$
(iv)

Therefore, coefficient of velocity,

$$C_v = v_{actual} / v_{theoretical}$$

$$= (\sqrt{gx^2/2y}) / (\sqrt{2Gh}) = \sqrt{x^2/4yH}$$
(v)

The simplest method of determining the coefficient of discharge (C_d) is by measuring the actual quantity of discharge through the orifice in a given time t. This actual discharge may then be divided by the theoretical discharge, which will give the required coefficient of discharge. Mathematically, coefficient of discharge is:

$$C_d = Q_{actual} / Q_{theoretical}$$
$$= Q_{actual}/\text{Area of orifice} \times \sqrt{2gH}$$

The coefficient of contraction may be found out by measuring the area of the jet at vena contracta and then by dividing the same by the area of the jet. Mathematically, coefficient of contraction is:

$$C_c = \text{Area of jet at vena contracta} / \text{Area of the orifice}$$

Since, the measurement of area of jet at vena contracta is difficult, hence the coefficient of contraction (C_c) can be easily found from the relation,

$$C_c = C_d / C_v$$

5.4 Discharge Through a Large Rectangular Orifice

The velocity of a liquid flowing through an orifice varies with the available head of that liquid. Hence, if the liquid is flowing through a large orifice, the velocity of the liquid particles will not be constant because there is a considerable variation of head along the height of the orifice. Usually, an orifice is considered to be large, if the liquid flow is less than 5 times the height of the orifice.

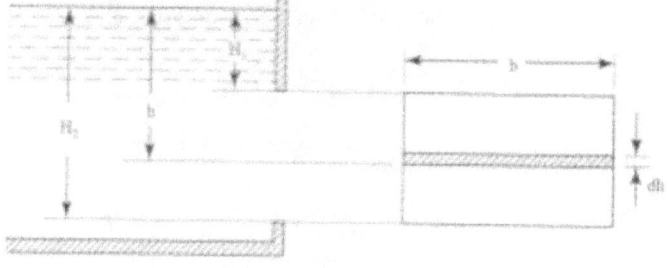

Fig (5-5) Rectangular orifice in one side of the tank

Now consider a large rectangular orifice, in one side of the tank, discharging water as shown in Figure (5-5).

Let,

H_1 = height of the liquid above the top of the orifice,

H_2 = height of liquid above the bottom of the orifice, and

b   = breadth of the orifice.

Consider a horizontal strip of thickness dh at a depth h from the water level as shown in Figure (5 - 5):

Therefore,   Area of strip = b   (i)

The theoretical velocity of water through the strip = $\sqrt{2gh}$ (ii)

Let,

dq = discharge through the strip,

C_d = coefficient of discharge, and

dq = C_d x Area x theoretical velocity = C_d b d h $\sqrt{2gh}$ (iii)

Total discharge through the whole orifice is found out by integrating the above equation between the limits H_1 and H_2:

$$Q = \int_{H_1}^{H_2} C_d \, b \, dh \, \sqrt{2gh}$$

$$= C_d \, b \, \sqrt{2g} \int_{H_1}^{H_2} \sqrt{h} \, dh$$

$$= 2/3 \, C_d \, b \, \sqrt{2g} \left[h^{3/2} \right]_{H_1}^{H_2}$$

$$= 2/3 \, C_d \, b \, \sqrt{2g} \, (H_2^{3/2} - H_1^{3/2})$$

5.4.1 Submerged or drowned orifice: Cases when an orifice does not discharge liquid freely into the atmosphere, but discharges into some other vessel containing liquid. Such an orifice is known as submerged or drown orifice. The two types of submerged orifices are:

(1) Wholly drowned orifice, and

(2) Partially drowned orifice.

5.4.2 Discharge through a wholly drowned orifice: If the outlet side of the orifice is completely submerged under the liquid surface, it is the known as 'wholly drowned or wholly submerged' orifice, as demonstrated in Figure (5- 6):

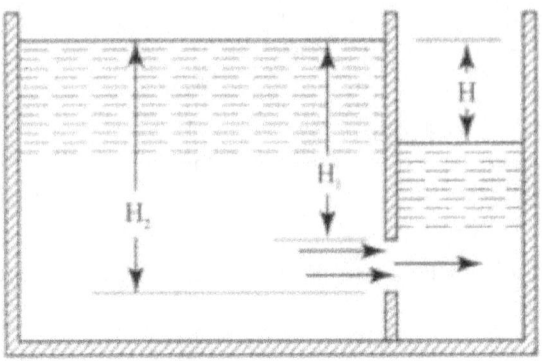

Fig (5-6) Wholly drowned orifice

Consider a wholly drowned orifice discharging liquid as shown in Figure (5 - 6):
Let,
H_1 = height of water (on the upstream side) above the top of the orifice,
H_2 = height of water (on the upstream side) above the bottom of the orifice,
H = difference between the two liquid levels on either side of the orifice,
b = width of the orifice, and
C_d = coefficient of discharge.

Area of the orifice = $b (H_2 - H_1)$
The theoretical velocity of water through an orifice = $\sqrt{2gh}$

Therefore, Actual velocity of water = $C_d\sqrt{2gh}$

Discharge through the orifice, Q = Area of orifice x actual velocity
$$= b (H_2 - H_1) \times C_d\sqrt{2gh}$$
$$= C_d \, b \, (H_2 - H_1) \times C_d\sqrt{2gh}$$

If the depth of the drowned orifice (d) is given instead of H_1 and H_2, then the discharge through a wholly drowned orifice is given by the relation:

$$Q = C_d \, b \, d \, \sqrt{2gh}$$

5.4.3 Discharge through a partially drowned orifice: If the outlet side of the orifice is partly underwater, it is known as a partially drowned or partly submerged orifice, as shown in Fig (5-7):

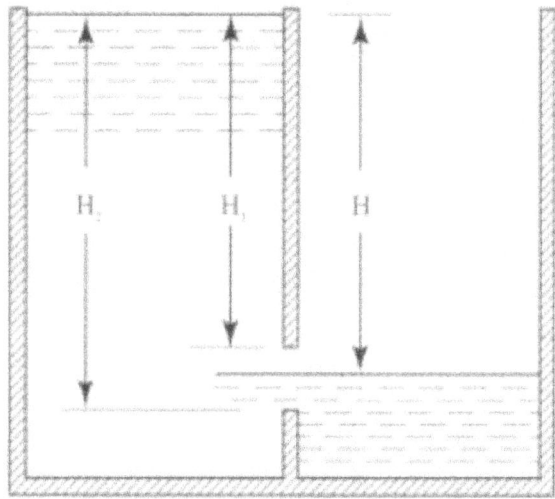

Fig (5-7) Partially drowned orifice

The discharge through a partially drowned orifice is obtained by treating the lower portion as a drowned orifice and the upper portion as an orifice running free. By adding the two discharges we thus obtain:

Discharge through the drowned portion:
$$Q_1 = C_d \, b \, (H_2 - H_1) \times \sqrt{2gh}$$

Discharge through the free portion:
$$Q_2 = 2/3 \, C_d \, b \, \sqrt{2gh} \, (H_2^{3/2} - H_1^{3/2})$$

Total discharge:
$$Q = Q_1 + Q_2$$

5.5 Mouthpieces

It was only after a series of experiments, that engineers found out that a short pipe fitted to an orifice will increase the value of coefficient of discharge. The increase in value of coefficient of discharge will increase the discharge through the orifice. Such a pipe, whose length is generally more than 2 times the diameter of the orifice, and is fitted (externally or internally) to the orifice, is known as 'mouthpiece'.

5.5.1 Types of mouthpieces: There are many types of mouthpieces, depending upon their size, shape or nature of discharge. The following types are important from the subject point of view:

(1) Related to position of the mouthpiece:
      (a) Internal mouthpiece, and
      (b) External mouthpiece.

(2) Related to shape of the mouthpiece:
      (a) Cylindrical mouthpiece,
      (b) Convergent mouthpiece, and
      (c) Convergent - divergent mouthpiece.

(3) Related to nature of discharge:
      (a) Mouthpiece running full, and
      (b) Mouthpiece running free.

5.5.2 Loss of head of liquid flowing in a pipe: Liquid flowing in a pipe loses its energy (or head) due to friction between pipe walls, change in section or obstruction.

All such losses are expressed in terms of velocity head. There are many type of losses of head, yet the following losses are important from the subject point of view:
(1) Loss of head due to sudden enlargement,
(2) Loss of head due to sudden contraction,
(3) Loss of head at entrance of a pipe,
(4) Loss of head at exit of a pipe, and
(5) Loss of head due to obstruction in a pipe.

5.5.3 Loss of head due to sudden enlargement: Consider a liquid flowing in a pipe ABC having a sudden enlargement at B. Eddies will be formed in corners as a result of such sudden enlargement, as shown in Figure (5-8):

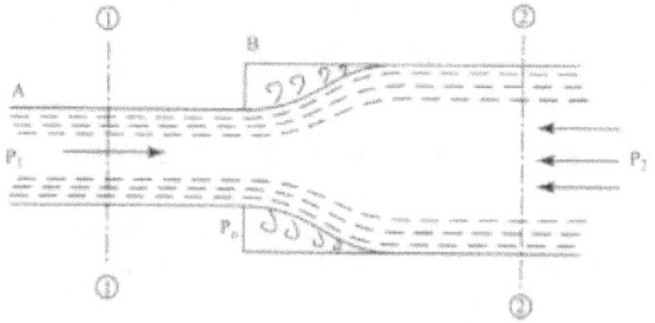

Fig (5-8) Pipe with sudden enlargement

Let,

a_1 = area of pipe at section 1-1,

v_1 = velocity of the liquid at section 1-1,

p_1 = pressure of the liquid at section 1-1,

a_2 = area of pipe at section 2-2,

v_2 = velocity of the liquid at section 2-2,

p_2 = pressure of the liquid at section 2-2,

p_0 = pressure of the liquid eddies on area ($a_2 - a_1$), and

h_e = loss of head due to sudden enlargement.

It is found that p_1 and p_0 is nearly equal. The head losses due to eddies take place at the sudden enlargement section as shown in Figure (5 - 8):

Applying Bernoulli's equation to sections 1-1 and 2-2:

$$v_1^2/2g + p_1/\omega = v_2^2/2g + p_2/\omega + losses \quad (taking\ Z_1 = Z_2)$$

Or losses, $\qquad h_e = (p_1/\omega - p_2/\omega) + (v_1^2/2g - v_2^2/2g)$ $\qquad\qquad$ (i)

The concentration of water/sec, at section 1-1:

$\qquad$ Mass x velocity = $(wa_1 v_1 \times v_1)/g = (wa_1\ v_1^2)/g$

Similarly, momentum of water/sec, at section 2-2:

$\qquad\qquad = wa_1\ v_1^2/g$

Therefore,

Change of momentum/sec = $(wa_1\ v_1^2)/g - (wa_2\ v_2^2)/g$ $\qquad\qquad$ (ii)

The flow between sections 1-1 and 2-2 is continuous,

Therefore,

$$a_1 v_1 = a_2 v_2$$
$$a_1 = (a_2 v_2)/v_1$$

Substituting this value of a_1 in equation (ii):

Change of momentum/sec = $(wv_1^2/g \times (a_2 v_2)/v_1 - (wa_2\ v_2^2)/g$

$\qquad\qquad = (wa_2 v_2 \times v_1)/g - (wa_2\ v_2^2)/g$

The force responsible for this change of momentum/sec = $p_2 a_2 - p_1 a_1 - p_0 (a_2 - a_1)$ =

$\qquad\qquad\qquad\qquad = p_2 a_2 - p_1 a_1 - p_0 a_2 + p_0 a_1$

$\qquad\qquad\qquad\qquad = a_2 (p_2 - p_1) \quad (taking\ p_1 = p_2)\ (iii)$

Equating (ii) and (iii):

$\qquad\qquad a_2 (p_2 - p_1) = (wa_2 v_2 \times v_1)/g - (wa_2\ v_2^2)/g$

$$= w\, a_2\, [(\upsilon_2 \times \upsilon_1)\,/\,g - (\upsilon_2^2)\,/g]$$
$$p_2/\omega - p_1/\omega = (\upsilon_1 \times \upsilon_2)\,/\,g - (\upsilon_2^2)\,/g$$
Or, $$\qquad p_1/\omega - p_2/\omega = (\upsilon_2^2)\,/g - (\upsilon_2 \times \upsilon_1)\,/\,g$$

Substituting this value of $p_1/\omega - p_2/\omega$ in equation (i):

Losses,

$$h_e = (\upsilon_2^2)\,/g - (\upsilon_2 \times \upsilon_1)\,/\,g + (\upsilon_1^2)/2g - (\upsilon_2^2)/2g$$
$$= 1/2g\,(2\upsilon_2^2 - 2\,\upsilon_1\upsilon_2 + \upsilon_1^2 - \upsilon_2^2)$$
$$= 1/2g\,(\upsilon_1^2 - 2\,\upsilon_1\upsilon_2 + \upsilon_2^2) = (\upsilon_1 - \upsilon_2)^2/2g$$

5.5.4 Loss of head due to sudden contraction:

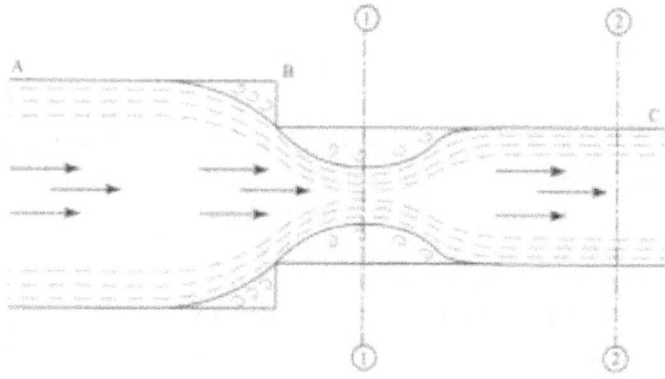

Fig (5-9) Pipe with sudden contraction

The loss of head due to sudden contraction is not due to contraction itself, but it is due to sudden enlargement, which follows the contraction.

Consider a liquid flowing in a pipe ABC, having a sudden contraction at B, As shown in Figure (5-9):

Let,

a_1 = area of pipe at section 1-1,

υ_1 = velocity of the liquid at section 1-1,

a_2 = area of pipe at section 2-2,

υ_2 = velocity of the liquid at section 2-2,

h_e = loss of head due to sudden contraction.

Since, the loss of head due to sudden enlargement is:

$$h_e = (\upsilon_1 - \upsilon_2)^2/2g \qquad\qquad\qquad (i)$$

And,

$$a_1\upsilon_1 = a_2\upsilon_2 \qquad\qquad\qquad (ii)$$

As,

$$a_1/a_2 = 0.62 \quad \text{(where, } 0.62 = C_c = \text{coefficient of contraction)}$$

Or,

$$a_1 = 0.62 \, a_2$$

Substituting value of a_1 in equation (ii):

$$0.62 \, a_2 v_1 = a_2 v_2$$

Therefore,

$$v_1 = v_2 / 0.62 \qquad \qquad \text{(iii)}$$

The liquid, while flowing through a narrow pipe, will be further contracted at section 1-1 forming a 'vena contracta' (in the same way as a jet issuing from an orifice). Therefore, the loss of head due to sudden contraction will actually be due to sudden enlargement from section 1-1 to section 2-2.

Loss of head due to sudden enlargement after section 1-1:

$$h_e = (v_1 - v_2)^2/2g$$

Substituting the value of v_1 from equation (iii):

$$h_e = (v_2/0.62 - v_2)^2/2g = 0.375 \, v_2^2/2g = K \, v_2^2/2g$$

Remarks:

1) The above relation is derived on the assumption that the value of the coefficient of contraction is 0.62. As the value of coefficient of contraction depends on the type of orifice used, consequently, the exact relation will vary with type of orifice.

2) By experiment it was found that the actual value of loss of head depends on the ratio of the two diameters i.e. d_1/d_2, where d_1 is the pipe's diameter at section 1 and d_2 is its diameter at section 2. The following table gives the values of k for corresponding values of d_1/d_2:

d_1/d_2	1.0	1.1	1.25	1.5	2.0	2.5	3.0	3.5	4.0
K	0	0.1	0.19	0.28	0.375	0.40	0.42	0.43	0.45

3) Although some references will differ on the value of K, yet it is internationally accepted to be as 0.375

5.5.5 Loss of head at pipe's entrance: The loss of head at entrance to the pipe is a loss due to sudden contraction and depends upon the form of entrance. The value of loss of head at entrance was found to be equal to 0.5 ($v_2^2/2g$), where v is the velocity of liquid flowing inside the pipe.

With long pipes, this loss of head is of small value when compared to frictional losses and thus neglected.

5.5.6 Loss of head at pipe exit: This loss of head, due to pipe exit, is a loss due to energy of head by virtue of liquid's motion. The value of head loss at exit was found to be equal to $v_2^2/2g$ where v is the velocity of liquid in the pipe.

In case of long pipes, the head loss, similar to entrance, is of a small value when compared to frictional losses and thus neglected.

5.5.7 Loss of head due to obstruction in a pipe:

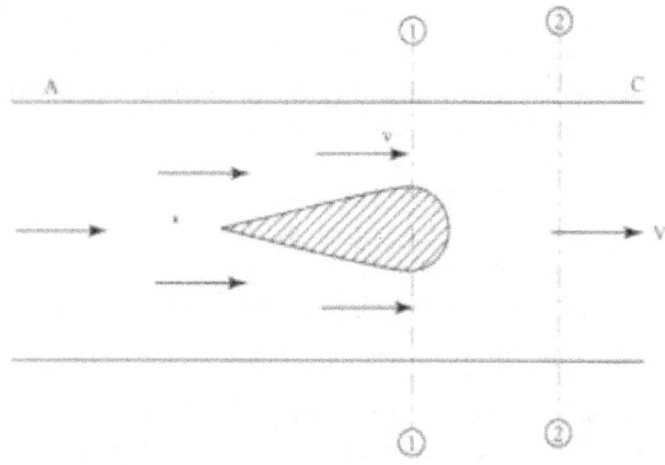

Fig (5-10) Liquid flowing in a pipe with an obstruction

This is a particular case of loss of head due to sudden contraction, which further depends on losses of head due to sudden contraction, and sudden enlargement.

Consider liquid flowing in a pipe having an obstruction as shown in Figure (5-10):
Let,
A = area of pipe at section 2-2,
a = area of obstruction,
Therefore,
(A -a) = area of pipe at section 1-1 through which the liquid has to pass,
V = velocity of liquid at section 2-2, and
v = velocity of liquid at section 1-1.

$$A \times V = (A - a) \times v \times C_c$$

Where,
C_c = coefficient of contraction
Therefore,

$$\upsilon = (A \times V) / [(A - a) \times C_c] \tag{i}$$

The loss of head due to sudden enlargement is:

$$(\upsilon_1 - \upsilon_2)^2/2g \tag{ii}$$

The liquid while flowing through the obstruction will further contract (in the same way as jet issuing from an orifice). Therefore, the loss of head will actually be due to sudden enlargement from section 1-1 to section 2-2:

Loss,

$$h_0 = (\upsilon - V)^2 / 2g$$

Substituting the value of υ from equation (i):

$$h_0 = \{(A \times V) / [(A - a) \times C_c] - V\}^2/2g$$
$$= V^2/2g \, [A/(A-a) \times C_c - 1]$$

5.6 Discharge through a Mouthpiece

If a mouthpiece is fitted to an orifice, it will increase the value of coefficient of discharge for the orifice. The increase in the value of coefficient of discharge will increase the discharge the discharge through the orifice.

5.6.1 Discharge through external mouthpiece: The discharge through an orifice may be increased by fitting a sufficient length of pipe to the outside of the orifice as shown in Figure (5-11). Such a pipe, which is attached externally to an orifice, in known as 'external mouthpiece':

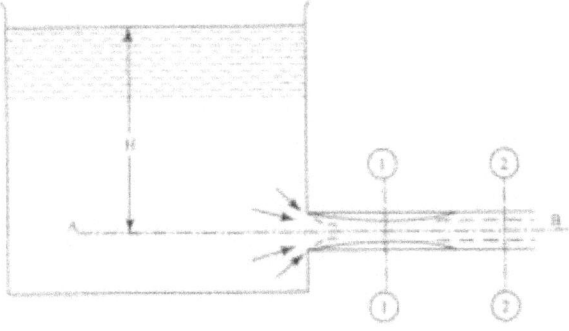

Fig (5-11) Tank with length of pipe to the outside of the orifice

The jet at pipe entry will first contract, then expand and fill the whole pipe as shown in Fig (5-11).

Let,

H = height of liquid above the mouthpiece,

a = area of orifice or mouthpiece,

a_c = area of flow at vena contracta,

v = velocity of liquid at outlet, and

v_c = velocity of liquid at vena contracta.

Assuming the coefficient of contraction C_c at vena contracta to be 0.62, we get:

$$a_c = C_c \times a = 0.62 \, a$$

As the liquid is continuously flowing, therefore form the equation of continuity we get:

$$v_c \, a_c = v \, a$$

Or,

$$v_c = v \, a / a_c = v / 0.62 \qquad\qquad (a/a_c = 0.62)$$

The jet passing through section 1-1 will enlarge at section 2-2. Therefore there is loss of head due to flow's sudden enlargement:

Therefore,

$$h_e = (v_c - v)^2/2g = (v/0.62 - v)^2/2g = 0.375 \, v^2/2g \qquad\qquad (i)$$

Applying Bernoulli's equation to points A and B:

$$H = v^2/2g + \text{losses} = v^2/2g + 0.375 \, v^2/2g = 1.375 \, v^2/2g$$

Or,

$$v^2/2g = H / 1.375$$

$$v = \sqrt{2gH / 1.375} = 0.855 \sqrt{2gH} \qquad\qquad (ii)$$

Since, velocity of liquid at outlet = $\sqrt{2gH}$

Therefore, coefficient of velocity:

$$C_v = \text{Actual velocity/Theoretical velocity}$$
$$= [0.855\sqrt{2gH}] / \sqrt{2gH} = 0.855 \qquad\qquad (iii)$$

To find out the discharge through an external mouthpiece, the value of coefficient of discharge for the mouthpiece is first determined as follows:

$$C_d = \text{Coefficient of velocity x coefficient of contraction}$$
$$= 0.855 \times 1 = 0.855$$

The above shows that the coefficient of discharge has considerably increased by fitting an external mouthpiece. The increase in coefficient of discharge, means the increase in the discharge through the orifice.
Therefore,

$$\text{Discharge, } Q = C_d \, a \, \sqrt{2gH} = 0.855 \, a \, \sqrt{2gH} \qquad (\text{as } C_d = 0.855)$$

It was experimentally found that there is some loss of head at the entrance of the mouthpiece. This loss depends on the type of orifice. Sometimes the loss of head reduces the coefficient of discharge by a small amount (up to 0.82); but for all practical purposes the value of coefficient of discharge is taken as 0.855

5.6.2 Discharge through internal mouthpiece: An internal mouthpiece, extending into the fluid i.e. inside the vessel, is also known as 'Re-entrant or Borda's mouthpiece'. Following are two types of internal mouthpieces depending upon their nature of discharge:
(a) Mouthpiece running free, and
(b) Mouthpiece running full.

If the jet, after contraction, does not touch the sides of the mouthpiece, it is said to be running free. But if the jet expands and fills up the whole mouthpiece, it is said to be running full. It is found that if the length of the mouthpiece is less than 3 times the diameter of the orifice, it will run free. But if the length of the mouthpiece is more than 3 times the diameter of the orifice, it will run full. The coefficient of discharge will be different in both cases.

5.6.3 Mouthpiece running free: Consider a mouthpiece running free as shown in Figure (5-12) below.
Let,
H = height of the liquid above the mouthpiece,
a = area of orifice or mouthpiece,
a_c = area of contracted jet,
v = velocity of liquid, and
w = sp. weight of the liquid.

Therefore, pressure of liquid on the mouthpiece, p = w H

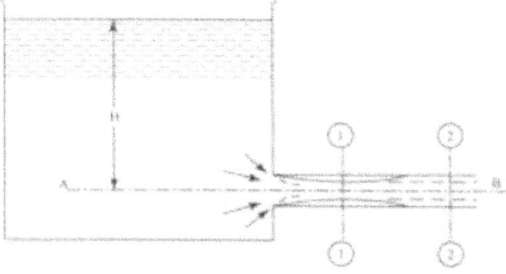

Fig (5-12) Tank with mouthpiece running free

And force acting on the mouthpiece = Pressure x area = w H a   (i)

Mass of the liquid flowing per second = $w\, a_c\, v/g$
Therefore,
Momentum of the flowing liquid/sec = mass x velocity = $w\, a_c\, v \times v/g = w\, a_c\, v^2/g$ (ii)

Since the water is initially at rest, therefore, the initial momentum = 0, and change of momentum = $w\, a_c\, v^2/g$

According to Newton's second law of motion (the rate of change of momentum is directly proportional to the acting force and takes place in the same direction in which the force is acting), the force is equal to the rate of change of momentum/sec i.e. equating (i) and (ii):
$$w\, H\, a = w\, a_c\, v^2/g$$
Or,
$$H\, a = a_c\, v^2/g$$

Substituting $H = v^2/g$ in the above equation:
$$v^2\, a/2g = a_c\, v^2/g$$
$$a = 2\, a_c$$
$$a_c/a = \tfrac{1}{2} = 0.5$$

But a_c/a is the coefficient of contraction by definition:
Therefore,
$$C_c = 0.5$$

If the coefficient of velocity $C_v = 1$, therefore the coefficient of discharge is:

$$C_d = C_v \times C_c = 1 \times 0.5 = 0.5$$
Therefore,
$$\text{Discharge, } Q = C_d\, a\, \sqrt{2gH} = 0.5\, a\, \sqrt{2gH}$$

5.6.4 Mouthpiece running full: If the jet after contraction expands and fills up the whole mouthpiece, it is said to be running full.
Consider a mouthpiece running full as shown in Figure (5-13):

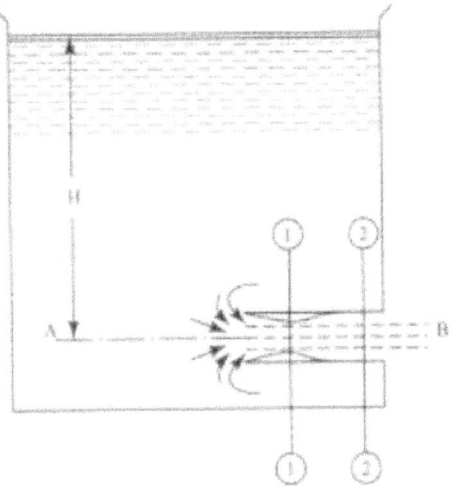

Fig (5-13) Mouthpiece running full

Let,

H = height of the liquid above the mouthpiece,

a = area of orifice or mouthpiece,

a_c= area of flow at vena contracta,

v = velocity of liquid at outlet, and

v_c= velocity of the liquid at vena contracta.

From continuity equation, as the liquid is flowing continuously, we get:

$$v_c \, a_c = v \, a$$

Therefore,

$$v_c = v \, a \, / \, a_c$$

The coefficient of contraction, in case of an internal mouthpiece, is 0.5. Substituting the value of $a_c/a = 0.5$ (or $a/a_c = 2$) in the above equation:

$$v_c = 2 \, v \qquad\qquad \text{(i)}$$

Jet passing through section 1-1 enlarges at section 2-2. Hence, there will be loss of head due to sudden enlargement, or:

$$h_e = (v_c - v)^{\,2}/2g = (2v - v)^2 \, / \, 2g \qquad\qquad (\text{as } v_c = 2 \, v)$$

Applying Bernoulli's equation to points A and B:

$$H = v^2/2g + losses = v^2/2g + v^2/2g = 2 \, v^2/2g = v^2/g$$

Or,

$$\upsilon^2 = g\,H$$
$$\upsilon = \sqrt{gH}$$

Therefore,

$$\text{Actual discharge} = a \times \sqrt{gH}$$

Since,

$$\text{The theoretical discharge} = a\,\sqrt{2gH}$$

Therefore,

$$\text{Coefficient of discharge, } C_d = \text{Actual discharge / Theoretical discharge}$$
$$= a\,\sqrt{gH}\,/\,a\,\sqrt{2gH} = 1/\sqrt{2} = 0.707$$

Therefore,

$$\text{Discharge, } Q = C_d\,a\,\sqrt{2gH} = 0.707\,a\,\sqrt{2gH}$$

It was found experimentally that the actual value of coefficient of discharge is slightly more than 0.707. Nevertheless, for practical reasons, the value of coefficient of discharge is taken as 0.707.

It was demonstrated that in case of internal mouthpiece, the coefficient of discharge is less than that of the external mouthpiece. The reason for this is that in case of external mouthpiece the liquid particles will deviate through a maximum angle of 90°.inern

With internal mouthpiece, the liquid particles will deviate through a maximum angle of 180°. Because of the larger angle of deviation of the liquid particles, the jet contraction is more in the case of internal mouthpiece than that in the external mouthpiece.

5.6.5 Discharge through a convergent mouthpiece:

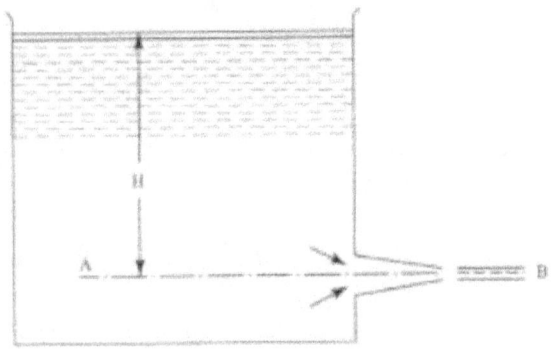

Fig (5-14) Mouthpiece with same shape as that of jet

As the jet of liquid contracts at the entrance of a mouthpiece, there is always a loss of head. To counteract this loss of head, due to contraction of the jet up to vena contracta, the mouthpiece is given the same slope as that of the jet of liquid, as shown in Figure (5-14).

The mouthpiece shown has no loss of head due to expansion. The mouthpiece ending at the vena contracta of the jet is called 'convergent mouthpiece'.
Let,
H = height of the liquid above the mouthpiece,
a = area of orifice at point B, and
v = velocity of jet.

Applying Bernoulli's equation to points A and B:

$$H = v^2/g \qquad \text{(as there is no loss of head)}$$

Or,

$$v = \sqrt{2gH}$$

Therefore,

Actual discharge, Q = Area of orifice x actual velocity

$$= a\sqrt{2gH} \qquad \text{(i)}$$

= coefficient of discharge x area of orifice x theoretical velocity

$$= C_d\, a\sqrt{2gH} \qquad \text{(ii)}$$

Equating (i) and (ii):

$$a\sqrt{2gH} = C_d\, a\sqrt{2gH}$$

Therefore,

Coefficient of discharge, $C_d = a\sqrt{2gH} \,/\, a\sqrt{2gH} = 1$

Therefore,

Discharge, $Q = a\sqrt{2gH}$

As there is a loss of head at the entrance of the mouthpiece, it will have the effect of slightly reducing the coefficient of discharge (sometimes up to 0.975). But for all practical purposes, the value of coefficient of discharge is taken as 1.

5.6.6 Discharge through a convergent-divergent mouthpiece (or Bell-mouthpiece): The mouthpiece is first made as a convergent shape up to the vena contracta of the jet, and beyond that, is made as a divergent shape. Such a mouthpiece is known as 'convergent-divergent mouthpiece' as shown in Figure (5-15).

There is no loss of head due to sudden expansion in this type of mouthpiece. In this case, the value of coefficient of discharge (derived similar to convergent mouthpiece) is also equal to1. The diameter of the mouthpiece, for the purposes of calculating discharge, is taken at vena contracta i.e. at point B (where the convergent and divergent pieces meet). It is also known as 'throat diameter' of the mouthpiece.

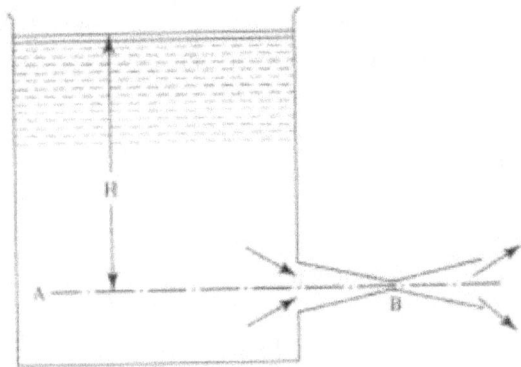

Fig (5-15) Convergent-divergent mouthpiece

5.6.7 Flow through nozzles:

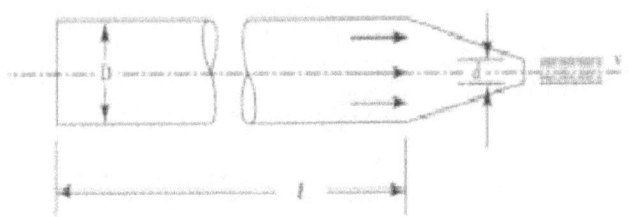

Fig (5-16) Flow through a nozzle

A nozzle is a tapering mouthpiece fitted to the outlet end of a pipe. It is used in achieving high velocity of fluid, as it converts pressure head into kinetic head at its outlet. A high velocity of water is required in firefighting, mining and power developments.

Consider a nozzle fitted at the outlet of a pipeline as shown in Figure (5-16):

Let,

L = length of pipe,

D = diameter of the pipe,

Area of pipe, A = π/4 (D)^2

V = velocity of water in the pipe,

f = coefficient of friction,

d = diameter of nozzle,

Area of nozzle, a = π/4 (d)^2

v = velocity of water through the nozzle, and

H = head of water, under which the flow takes place.

As the water is continuously flowing through the pipe nozzle, therefore:

AV = a v, and
V = α v / A

Darcy's formula for loss of head in pipes due to friction = $(4 f L V^2) / (2 g D)$ (i)
Loss of head due to velocity at outlet = $v^2/2g$ (ii)
Hence,
H = loss of head due to friction + loss of head due to velocity at outlet
$\quad\quad$ = $(4 f L V^2) / (2 g D) + v^2/2g$ (neglecting minor losses)
$\quad\quad$ = $v^2/2g (4 f L/D \times a^2/A^2 + 1)$
$\quad\quad$ = $4 f L/2 g D (a^2v^2/ A^2) + v^2/2g$
Therefore,
$$v = \sqrt{(2gH)/ (1 + 4 f L/D \times a^2/A^2}$$ (iii)
And,
$\quad\quad$ $v^2/2g = H - (4 f L V^2 / 2gD)$
$\quad\quad\quad$ = $H - [4 f L / 2gD (a^2v^2/ A^2)]$ (iv)

The kinetic energy of jet through the nozzle = w Q $v^2/2g$ m-kg/sec
Therefore,
Power available at the outlet of the jet = $(w Q v^2/2g) / 75$ = $[w (a v) v^2/2g] / 75$
$\quad\quad\quad\quad\quad\quad\quad\quad\quad$ = $w a v / 75 \times v^2/2g$ HP   (as Q = a v)
$\quad\quad\quad\quad\quad\quad\quad\quad\quad$ = $w a v / 75 [H - 4 f L/ 2 g D (a^2v^2/ A^2)$ HP   (v)

Power available at inlet end of the pipe = w Q H / 75
Therefore,
Efficiency of transmission, η = power available at outlet of pipe / power available at inlet of
pipe = $[w Q v^2/2g \times 75] / [w Q H / 75]$ = $v^2/2g$ H

The power transmitted will be maximum when d(HP) /d v = 0, or when the differential
coefficient of the amount inside brackets of equation (v) is zero:
$\quad\quad\quad\quad$ d{ w a v / 75 [H - 4 f L/ 2 g D (a^2v^2/ A^2)]} / dv = 0
Or,$\quad\quad\quad\quad$ d{ w a / 75 [H - 4 f L/ 2 g D (a^2v^3/ A^2)]} / dv = 0
$\quad\quad\quad\quad$ H- 3[(4 f L/ 2 g D) x (a^2v^3/ A^2)] = 0
$\quad\quad\quad\quad$ H- 3[(4 f LV^2 / 2 g D) = 0   (as V = a v /A)
$\quad\quad\quad\quad$ H - 3 h_f = 0   (as h_f = 4 f L V^2/ 2gD)
Therefore,
$\quad\quad\quad\quad$ $h_f = H / 3$

The power transmitted through the nozzle is a maximum when the head lost due to friction in
the pipe is equal to 1/3 of the total supply head.

5.6.8 Diameter of the nozzle for maximum transmission of power: Consider a pipe, having a nozzle fitted at its outlet.

Let,

L = length of pipe,

D = diameter of pipe,

Therefore, area of pipe, $A = \pi/4 \ (D)^2$

V = velocity of water in pipe,

f = coefficient of friction,

d = diameter of nozzle

Therefore, area of nozzle, $a = \pi/4 \ (d)^2$

υ = velocity of water through nozzle, and

H = head of water

The power transmitted through the nozzle:
$$P = w \ a \ \upsilon/75 \ x \ \upsilon^2/2g = [w \ x \ \pi/4 \ (d)^2 \ x \ \upsilon] \ /75 \ x \ \upsilon^2/2g$$
$$= w/75 \ x \ \pi/4(d)^2 \ x \ \frac{1}{2} \ g \ x \ \upsilon^3 \tag{i}$$

The velocity of water through the nozzle:
$$\upsilon = \sqrt{2gH/(L + 4fL/D \ x \ a^2/A^2)}$$

Substituting the value of υ in equation (i):

$P = w/75 \ x \ \pi/4(d)^2 \ x \ L/2g \ [2g \ H/ \ (L + 4f \ L/D \ x \ a^2/A^2)]^{3/2}$

$\quad = w/75 \ x \ \pi/4(d)^2 \ x \ L/2g \ [L + 4f \ L/D) \ (\pi/4((d^2)^2/ \ \pi/4 \ ((D^2)^2]^{3/2}$

$\quad = w/75 \ x \ \pi/4(d)^2 \ x \ L/2g \ [2g \ H/(L + 4f \ L \ d^4/D^5]^{3/2}$

$\quad = w/75 \ x \ \pi/4(d)^2 \ x \ L/2g \ [2g \ H/(D^5 + 4f \ L \ d^4/D^5]^{3/2}$

$\quad = w/75 \ x \ \pi/4(d)^2 \ x \ L/2g \ [2g \ H \ D^5/ \ (D^5 + 4f \ L \ d^4/D^5]^{3/2}$

$\quad = w/75 \ x \ \pi/4 \ x \ 2g \ (2Ghd^5)^{3/2} \ x \ d^2/ \ (D^5 + 4f \ L \ d^4)^{3/2}$

$\quad = K \ (d^2/ \ (D^5 + 4f \ L \ d^4)^{3/2} \tag{ii}$

Where, K is a constant $= w/75 \ x \ \pi/4 \ x \ L/2g \ (2gHd^5)^{3/2}$

The power transmitted will have a maximum value when $d(HP)/d(d) = 0$, or when the differential coefficient of equation (ii) is equal to zero:

$$d[K \ (d^2/ \ (D^5 + 4f \ L \ d^4)^{3/2}] \ / \ d(d) = 0$$

Or,

$\{[(d^2/ \ (D^5 + 4f \ L \ d^4)^{3/2} \ x \ 2d] - [d^2 \ x \ 3/2 \ (D^5 + 4f \ L \ d^4)^{1/2} \ x \ 4 \ x \ 4 \ f \ Ld^3]\}/\{(D^5 + 4f \ L \ d^4)^{3/2}\} = 0$

$\{[(d^2/ \ (D^5 + 4f \ L \ d^4)^{3/2} \ x \ 2d] - [d^2 \ x \ 3/2 \ (D^5 + 4f \ L \ d^4)^{1/2} \ x \ 4 \ x \ 4 \ f \ Ld^3]\} = 0$

Therefore,

$[(d^2/ \ (D^5 + 4f \ L \ d^4)^{3/2} \ x \ 2d] - [(D^5 + 4f \ L \ d^4)^{1/2} \ x \ 24 \ f \ L \ d^5 = 0$

Or,

$(D^5 + 4f\,L\,d^4)^{3/2} \times 2d = [(D^5 + 4f\,L\,d^4)^{1/2} \times 24\,f\,L\,d^5(D^5 + 4f\,L\,d^4)^{3/2} \times 2d = 24\,f\,L\,d^5$

$D^5 + 4f\,L\,d^4 = 12\,f\,L\,d^4$

$8\,f\,L\,d^4 = D^5$

$d^4 = D^5/\,8\,f\,L$

$d = (D^5/\,8\,f\,L)^{1/4}$ (iii)

Equation (iii) can also be written as:

$D^5/d^4 = 8\,f\,L$

$D^4/d^4 = 8\,f\,L/D$

$D^2/d^2 = \sqrt{8\,f\,L/D}$ (taking square root)

Or,

$\pi/4\,(D^2)\,/\,\pi/4\,(d^2) = \sqrt{8\,f\,L/D}$ (multiplying and dividing by π/4)

$A/a = \sqrt{8\,f\,L/D}$

5.7 Reynolds Number Re

Reynolds found that the value of critical velocity is governed by the relationship between the inertia force and viscous forces. He derived a ratio of these two forces and found out a dimensionless number known as Reynolds number:

Re = Inertia forces / Viscous forces

$= \rho\,\upsilon^2/\,(\mu\,\upsilon/d) = \rho\,\upsilon\,d\,/\,\mu = \upsilon\,d\,/\,\upsilon$ (as υ = μ/ ρ)

= (Mean velocity of liquid x Diameter of pipe) / (Kinetic viscosity of liquid)

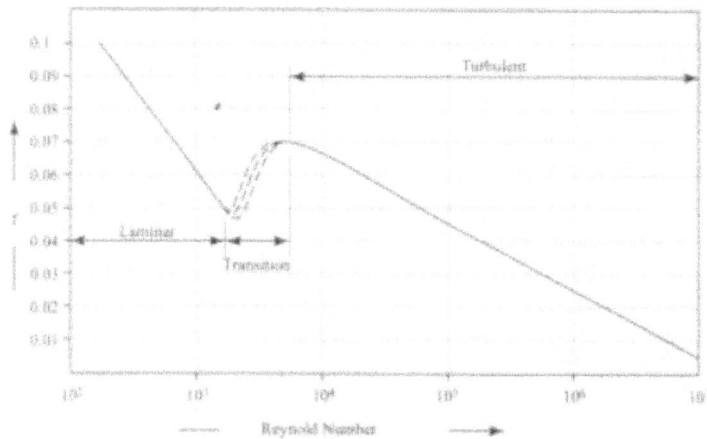

Fig (5-17) Reynolds number verses coefficient of friction

Reynolds number is a dimensionless quantity which has much importance and gives us the information about the types of flow (i.e. laminar or turbulent). Reynolds, after carrying out a series of experiments, found that if the Reynolds number for a particular flow is less than 2000, the flow is a laminar, between 2000 and 2800 it is neither laminar nor turbulent flow, and exceeding 2800 the flow is considered turbulent.

5.8 Weirs

A structure used to dam up a stream over a river, over which water flows, is called a weir. The conditions of flow of a weir are practically the same as those of a rectangular notch. That is way a notch is sometimes called a weir and vice versa.

The only difference between a notch and a weir is that the notch is of a smaller size than the weir. Also, a notch is usually made as a plate, whereas a weir is usually made of masonry or concrete.

5.8.1 Types of weirs: There are many types of weir depending on their shape, nature of discharge, width of crest or nature of crest. The following are important from the subject point of view:

1) According to shape,
   a) Rectangular weir, and
   b) Cippoletti weir.

2) According to nature of discharge,
   a) Ordinary weir, and
   b) Submerged or drowned weir.

3) According to the width of crest,
   a) Narrow-crested weir, and
   b) Broad-crested weir

4) According to the nature of crest,
   a) Sharp crested weir, and
   b) Ogee weir.

5.8.2 Discharge over a rectangular weir:

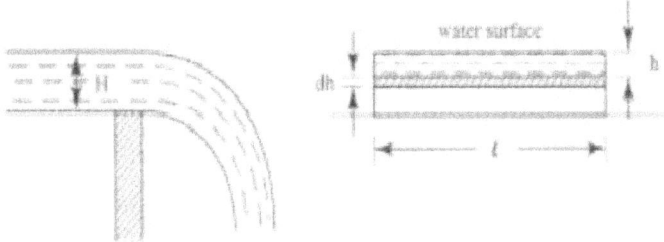

Fig (5-18) Rectangular weir

Consider a rectangular weir over which the water is flowing as shown in Figure (5-18):
Let,
H = height of the water above the crest of the weir, and
L = Length of the weir.

Considering a horizontal strip of water of thickness dh of a depth h from the water surface as shown in Figure (5-18):

$$\text{Area of the strip} = L \, dh \qquad\qquad\qquad\text{(i)}$$

The theoretical velocity of water through the strip = $\sqrt{2gh}$ $\qquad\qquad$ (ii)
Let,
dq = discharge through the strip, and
C_d = coefficient of discharge.

Therefore, dq = C_d x Area of strip x Theoretical velocity = C_d L dh $\sqrt{2gh}$ $\qquad$ (iii)

The total discharge, over the weir, may be found out by integrating the above equation within the limits O and H.

5.8.3 Francis formula for discharge over a rectangular weir (effect of end contractions):
Francis proposed an empirical formula for discharge over a rectangular weir. It was found that the length of the stream of liquid, while flowing over a weir, contracts at the ends of the still, as shown in Figure (5-19):

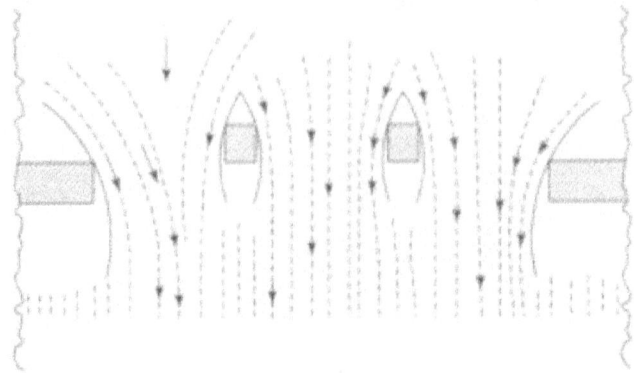

Fig (5-19) Stream of liquid flowing over a weir

This end contraction of the stream of liquid is known as 'lateral contraction' or 'side contraction'. Francis also found that the amount of the end contraction depends upon the conditions of sides of the channel, top of the still, and velocity of liquid. The value of end contraction, at each end, is approximately one-tenth the value of the height of liquid above the still of the weir.

If there are two contractions (as in case of a simple rectangular weir), the effective length of the weir is: (L - 0.2 H). Substituting this value of length in the equation for discharge, we get:

$$Q = 2/3 \ C_d(L - 0.2 \ H) \sqrt{2g} \ (H)^{3/2}$$

Sometimes, the total length of a weir is subdivided into a number of bays of spans by vertical posts, as shown in Figure (5-19). In such case, the number of end contractions will be twice the number of bays or spans into which the weir is divided. Thus in general, we can apply the empirical formula proposed by Francis as:

$$Q = 2/3 \ C_d \ (L - 0.1 \ n \ H) \sqrt{2g} \ (H)^{3/2}$$

Where,
n = number of end contractions.

Substituting $C_d = 0.623$ and $g = 9.81$ m/s^2, we get:

$$Q = 1.84 \ (L - 0.1 \ n \ H) \ H^{3/2}$$

When the end contractions are suppressed, the value of n in the above equation is taken as zero.

5.8.4 Basin's formula for discharge over rectangular weir: Basin proposed an empirical formula for the discharge over a rectangular weir. It was found that the value of coefficient of discharge varies with the height of the water over the still of a weir. Finally, he proposed an amendment to his discharge formula over a rectangular notch as below:

The discharge over a rectangular weir is:

$$Q = 2/3\ C_d\ L\ \sqrt{2g}\ (H)^{3/2}$$

Basin proposed:

$$Q = m\ L\ \sqrt{2g}\ (H)^{3/2}$$

Where, $m = 2/3\ C_d$

Basin also found that the value of m varies with the head of water, using the following relation:

$$m = 0.405 + (0.003/H)$$

where,
H = height of water, in meters

The above relations are correct by avoiding the effects of end contraction.

5.8.5 Discharge over a Cippoletti weir: The Cippoletti weir is a trapezoidal having side slopes of 1 horizontal to 4 vertical, as shown in Figure (5-20):

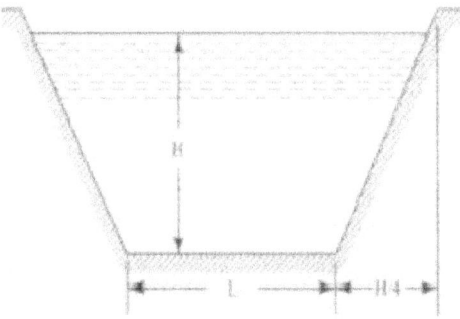

Fig (5-20) Cippoletti's weir

The slopes on the sides are given to obtain an increased discharge through the triangular portion of the weir, which would have been decreased due to end contractions in the case of rectangular weirs. Thus, the advantage of a Cippoletti weir is that the factor of end contracting is not required, while using the Francis formula.

A Cippoletti weir is a theoretical weir, whose side slope (1 horizontal to 4 vertical) is obtained by splitting the trapezoidal weir into a rectangular weir and a triangular notch. Since the discharge over a rectangular weir, considering end contraction is given by:

$$Q_1 = 2/3 \ C_d \ \sqrt{2g} \ (L - 0.2 \ H) \ (H)^{3/2} \qquad \text{(i)}$$

And discharge over the triangle notch is given by:

$$Q_2 = 8/15 \ C_d \ \sqrt{2g} \ \tan\Theta/2 \ (H)^{5/2} \qquad \text{(ii)}$$

Therefore the total discharge over the trapezoidal weir,

$$Q = Q_1 + Q_2$$
$$= 2/3 \ C_d \ \sqrt{2g} \ (L - 0.2 \ H) \ (H)^{3/2} + 8/15 \ C_d \ \sqrt{2g} \ \tan\Theta/2 \ (H)^{5/2} \qquad \text{(iii)}$$

Since the main idea of Cippoletti was to avoid the factor of end contraction, and as such he gave the formula for the discharge as:

$$Q = 2/3 \ C_d \ L \ \sqrt{2g} \ (H)^{3/2} \qquad \text{(iv)}$$

Equating equations (iii) and (iv), we get:

$$Q = 2/3 \ C_d \ L \ \sqrt{2g} \ (H)^{3/2} = 2/3 \ C_d \ \sqrt{2g} \ (L - 0.2 \ H) \ (H)^{3/2} + 8/15 \ C_d \ \sqrt{2g} \ \tan\Theta/2 \ (H)^{5/2}$$

Dividing the sides by $[2/3 \ C_d \ \sqrt{2g} \ (H)^{3/2}]$:

$$L = (L - 0.2 \ H) + \tan\Theta/2 \ (H) = L - 0.2H + 4/5 \ \tan\Theta/2 \ (H)$$

Or, $4/5 \ \tan\Theta/2 \ (H) = 0.2 \ H$

Or, $\tan\Theta/2 = 0.2 \times 5/4 = 1/5 \times 5/4 = \frac{1}{4}$

If a trapezoidal weir having side slopes, and 1 horizontal to 4 vertical (as is clear from the above equation i.e. $\tan\Theta/2 = \frac{1}{4}$) the factor of end contraction is not required for the discharge when using Francis formula. Since Cippoletti weir was designed on the above deviation, thus, the Francis formula for Cippoletti weir becomes:

$$Q = 1.84 \ L \ (H)^{3/2}$$

Or more accurately,

$$Q = 1.85 \ L \ (H)^{3/2} \qquad \text{(as proposed by Cippoletti)}$$

5.9 Velocity of Approach

Sometimes, a weir is provided in a stream or a river to measure the flow of water. In such case the water approaching the weir attains some velocity. Tis velocity is known as 'velocity of approach' and is assumed to be uniform over the whole weir.

In previous examinations, the velocity of approach was neglected and all formulae were derived on the assumption that the water on the upstream side of the weir is not in motion. By taking into consideration the value of velocity of approach which is affecting the discharge over the weir we have:

Let,

A = cross sectional area of the channel on the upstream side of the weir, and

Q = discharge over the weir.

The velocity of approach, $v = Q / A$ (i)

Since the formulae for discharge, over the weir, involves the height of water above the crest of the weir, this velocity of approach should also be converted into an additional head of water, acting over the whole weir.

Therefore, additional height of water, due to velocity of approach is:

$$H_a = v^2/2g$$

If the velocity of approach is considered for the discharge over the weir, then the additional height of water should also be taken into account.

Let,

H   = height of water, over the crest of the weir,

H_a = height of water, due to velocity of approach

Therefore, total height of water above the weir,

$$H_1 = H + H_a$$

It is thus obvious, that the limits of integration for the discharge over a rectangular weir will be H_a and H_1 instead of O and H. Therefore the discharge over the weir, with the velocity of approach is,

$$Q = 2/3 \ C_d \ L \ \sqrt{2g} \ [H^{3/2} - H_a{}^{3/2}]$$

And Francis formula for discharge with the velocity of approach,

$$Q = 1.84 \, (L - 0.1 \, n \, H_1) \, [H_1^{3/2} - H_a^{3/2}]$$

And Basin's formula for discharge with the velocity of approach,

$$Q = m \, L \, \sqrt{2g} \, (H_1)^{3/2}$$

5.9.1 Determination of velocity of approach: Sometimes the value of velocity of approach is not given. In such a case, the discharge over the weir is first found out, ignoring the velocity of approach. Then the velocity of approach is obtained by dividing the discharge by the cross-sectional area of the channel on the upstream side of the weir. This velocity of approach, so obtained, is then used for finding the discharge over the weir considering the velocity of approach.

If more accurate discharge is required, the above process is repeated. But it is generally observed that the discharge obtained by substituting the first obtained value, of the velocity of approach, does not differ appreciably from the further values. Thus the repetition is not necessary in practice.

5.9.2 Discharge over a submerged or drowned weir: When the water level on the downstream side of a weir is above the top surface of the weir, it is known as a submerged or drowned weir as shown in Figure (5-21):

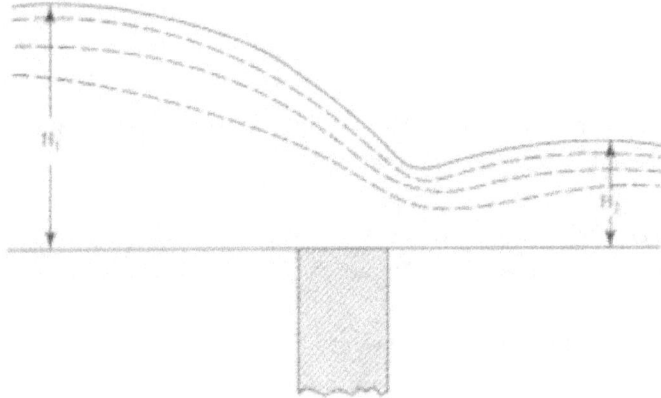

Fig (5-21) Submerged or drowned weir

The total discharge over the weir is found by splitting up the height of water, above the sill of the weir, into two portions as discussed below:

Let,

H_1 = height of water on the upstream side of the weir, and

H_2 = height of water on the downstream side of the weir.

The discharge over the upper portion may be considered as a free discharge under a head of water equal to (H_1 - H_2). The discharge over the lower portion may be considered as a submerged discharge under a head of H_2.

Thus, the free discharge over the upper portion is:

$$Q_1 = 2/3 \ C_d \ L \ \sqrt{2g} \ [H_1 - H_2]^{3/2}$$

And the submerged discharge over the lower portion is:

$$Q_2 = C_d \ L \ H \ \sqrt{2g \ (H_1 - H_2)}$$

Hence, the total discharge, $Q = Q_1 + Q_2$

5.9.3 Discharge over a sharp-crested weir: This is a special type of weir having sharp crest as shown in Figure (5-22):

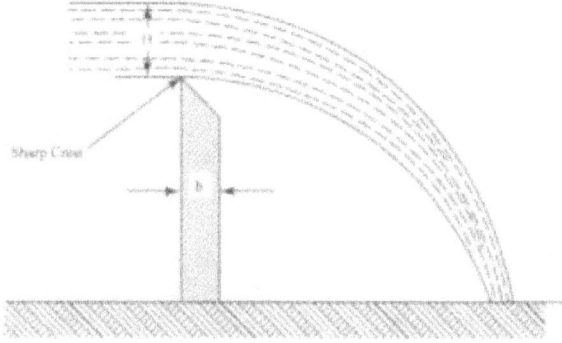

Fig (5-22) Sharp-crested weir

The water flowing over the crest comes in contact with the crest line, springs up and falls as a trajectory. In a sharp-crested weir its thickness is kept less than half the height of water on the weir:

$$b < H/2$$

Where,
b = thickness of the weir, and
H = height of water above the crest of the weir.

The discharge equation for a sharp crest weir remains the same as that of a rectangular weir:

$$Q = 2/3 \ C_d \ L \ \sqrt{2g} \ [H_1 - H_2]^{3/2}$$

Where,
C_d = coefficient of discharge, and
L = length of sharp-crested weir.

5.9.4 Discharge over an Ogee weir: It is a special type of weir, generally used as a spillway of a dam, as shown in Figure (5-23):

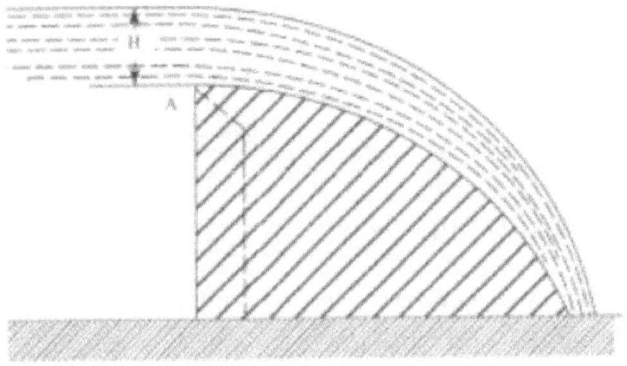

Fig (5-23) Ogee weir

The crest of an ogee weir is raised slightly above point A (i.e. crest of a sharp-crested weir) and on reaching the maximum rise of 0.115 H (where H is the height of water above the point A) falls in a parabolic form, as shown in Figure (5-23).

The discharge equation for an ogee weir remains the same as that of a regular weir:

$$Q = 2/3 \ C_d \ L \ \sqrt{2g} \ [H]^{3/2}$$

Where,
C_d = coefficient of discharge, and
L = length of an ogee weir.

5.9.5 Discharge over a triangular notch: Consider a triangular notch (or sometimes known as V-notch) placed on one side of the tank with water flowing over it as shown in Figure (5-24):

Let,
H = height of liquid above the apex of the notch, and
Θ = angle of the notch.

From figure's geometry the width of the notch is found to be = 2 H tanΘ/2

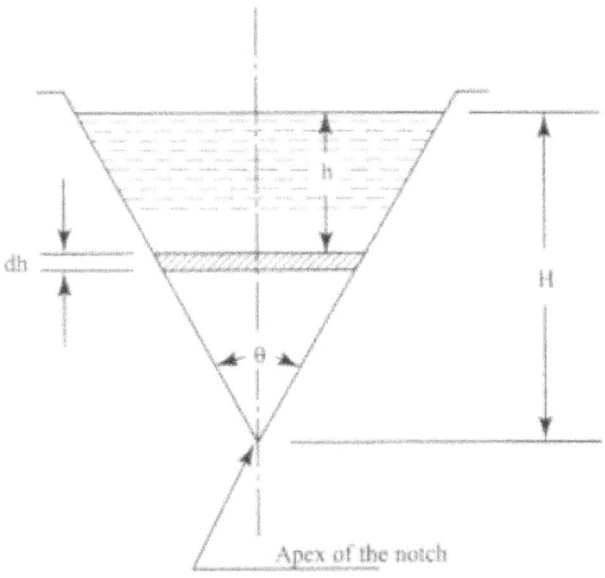

Fig (5-24) V-notch

Taking a horizontal strip of water of thickness dh and at a depth h from the water level, we get:

$$\text{Width of the strip} = 2\,(H - h)\,\tan\Theta/2$$

Therefore,

$$\text{Area of the strip} = dh\,2(H - h)\,\tan\Theta/2 \qquad (i)$$

$$\text{The theoretical velocity of water through the strip} = \sqrt{2gH} \qquad (ii)$$

Let,
dq = discharge through the strip, and
C_d = coefficient of discharge,

$$dq = C_d \times \text{Area of strip} \times \text{theoretical velocity}$$

$$dq = C_d\,dh\,2\,(H - h)\,\tan\Theta/2\,\sqrt{2gH} \qquad (iii)$$

The total discharge over the whole notch may be found out by integrating the above equation within limits of 0 and H:

$$Q = \int_0^H C_d\,dh\,2\,(H - h)\,\tan\Theta/2\,\sqrt{2gH}$$

$$= 2\,C_d\,\sqrt{2g}\,\tan\Theta/2\,\int_0^H (H - h)\,\sqrt{h}\,dh$$

$$= 2\ C_d\ \sqrt{2g}\ \tan\Theta/2 \int_0^H [(H\ h^{1/2}) - (h)^{3/2}]\ dh$$

$$= 2\ C_d\ \sqrt{2g}\ \tan\Theta/2\ [2/3\ H\ h^{3/2} - 2/5\ h^{5/2}]_0^H$$

$$= 8/15\ C_d\ \sqrt{2g}\ \tan\Theta/2\ (H)^{5/2}$$

When, $\Theta = 90°$, $C_d = 0.6$, and $g = 9.81$ m/sec^2, then:

$$Q = 1.417\ H^{5/2}$$

5.9.6 Advantage of a triangular notch over a rectangular notch: Following are advantages of using a triangular notch:

1) Only one reading i.e. head (H) is required for the measurement of discharge in a given triangular notch.

2) The formula for a triangular notch at $\Theta = 90°$ is simplified as $Q = 1.417\ H^{5/2}$.

3) A triangular notch gives more accurate results for low discharge than a regular notch.

4) The triangular notch can accurately measure a wide range of flow.

5.9.7 Time required in emptying a tank over a rectangular notch: Consider a tank of uniform cross-sectional area discharging liquid and having a rectangular notch in one of its vertical sides.
Let,
A = surface area of the tank,
H_1 = initial height of the liquid, above the sill of the notch,
H_2 = final height of the liquid, above the sill of the notch, and
T = time required in seconds, to lower the height of liquid from H_1 to H_2.

Let, at some instant, the height of liquid above the sill of the notch is h. A small quantity of liquid dq will flow over the notch at small intervals of time dt and reducing the liquid level by an amount dh in the tank.
Therefore,
$$dq = -\ A\ dh$$
(minus sign indicates that value of h will decrease as the discharge increases).

The discharge through the notch, dq = rate of discharge x time = $2/3\ C_d\ b\ \sqrt{2g}\ (H)^{3/2}$x dt

Or,
$$-A\ dh = 2/3\ C_d\ b\ \sqrt{2g}\ (H)^{3/2} x\ dt$$

Or,
$$dt = [-A\ dh] / [2/3\ C_d\ b\ \sqrt{2g}\ (h)^{3/2}]$$

The total time taken (in seconds) to lower the height of liquid from H_1 to H_2 may be found out by integrating the above equation between the limits H_1 to H_2:

$$T = \int_{H_1}^{H_2} [-A\ dh] / [2/3\ C_d\ b\ \sqrt{2g}\ (h)^{3/2}]$$
$$= (-A) / (2/3\ C_d\ b\ \sqrt{2g}) \int_{H_1}^{H_2} h^{-3/2} dh$$
$$= (-A) / (2/3\ C_d\ b\ \sqrt{2g}) [(h^{-1/2}) / (-1/2)]_{H_1}^{H_2}$$
$$= (2A) / (2/3\ C_d\ b\ \sqrt{2g}) [1/\sqrt{h}]_{H_1}^{H_2}$$
$$= (2A) / (2/3\ C_d\ b\ \sqrt{2g}) [1/\sqrt{h_2} - 1/\sqrt{h_1}]$$

5.9.8 Time of emptying a tank over a triangular notch: Consider a tank of uniform cross-sectional area and having a triangular notch in one of its sides discharging liquid:

Let,

A = surface area of tank,

Θ = angle of the notch,

H_1 = initial height of the liquid, above the apex of the notch,

H_2 = final height of the liquid, above the apex of the notch, and

T = time taken, in seconds, to reduce the height of liquid from H_1 to H_2 above the apex of the notch.

At some instant, the height of liquid above the apex of the notch is h. If a small quantity of liquid dq flows over the notch in small intervals of time dt, will then reduce the liquid level in tank by an amount of dh:

Therefore,

Discharge, $dq = -A\ dh$

(The minus sign indicate that the value of h will decrease as the discharge increase).

The discharge through the notch, dq = rate of discharge x time
$$= 8/15\ C_d\ \sqrt{2g}\ \tan(\Theta/2)\ (H)^{5/2} dt$$

Or,
$$-A\ dh = 8/15\ C_d\ \sqrt{2g}\ \tan(\Theta/2)\ (h)^{5/2} dt$$

Or,
$$dh = [-A\ dh] / [8/15\ C_d\ \sqrt{2g}\ \tan(\Theta/2)\ (h)^{5/2}]$$

The total time taken (in seconds) to lower the height of liquid from H_1 to H_2 may be found out by integrating the above equation between the limits H_1 to H_2:

$$T = \int_{H_1}^{H_2} [-A\, dh] \,/\, [8/15\ C_d\ \sqrt{2g}\ \tan(\Theta/2)\ (h)^{5/2}]$$

$$= [-A] \,/\, [8/15\ C_d\ \sqrt{2g}\ \tan(\Theta/2) \int_{H_1}^{H_2} (h)^{5/2}\, dh$$

$$= [-A] \,/\, [8/15\ C_d\ \sqrt{2g}\ \tan(\Theta/2)\ [(h^{-3/2})\,/\,(-3/2)]_{H_1}^{H_2}$$

$$= [-A] \,/\, [8/15\ C_d\ \sqrt{2g}\ \tan(\Theta/2)\ [1/h^{-3/2}]_{H_1}^{H_2}$$

$$= [-A] \,/\, [8/15\ C_d\ \sqrt{2g}\ \tan(\Theta/2)[1/\,H_2{}^{3/2} - 1/H_1{}^{3/2}]$$

5.10 Gauging of Rivers and Channels

People have intervened in the natural course and behaviour of rivers since before recorded history, to manage the water resources, to protect against flooding, or to make passage along or across rivers easier.

Stream discharge can be measured using:
(1) volumetric gauging,
(2) float gauging,
(3) current metering,
(4) chemical method (constant injection of chemical solution),
(5) structural methods, and
(6) slope-area methods

The choice of method depends on the characteristics of the stream and on the application.

5.10.1 Floats: A simple way of measuring the velocity of flow is by means of floats. The surface velocity, at any section, can be obtained with the help of a single float. It works by observing the time taken by the float to travel a known distance. The velocity is then calculated by dividing the distance the float travelled by the time taken to travel that distance. This surface velocity can then be converted into an average velocity.

A better method is to use a double floats or rod floats, which directly give the average velocity of flow:

a) Double floats:

A double float consists of two floats connected by a wire or a string as shown in Figure (5-25):

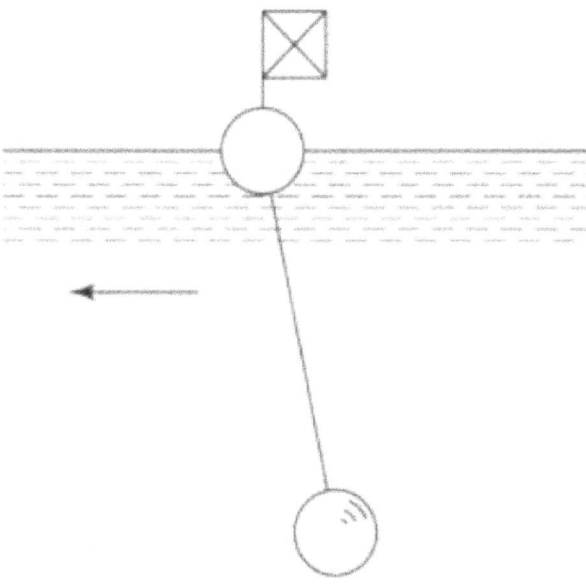

Fig (5-25) Double floats

One of the floats is a wooden-float, floating on the surface, while the other float is a hollow metallic sphere (heavier than water) and is suspended from the former by a wire or string of known length. As the average velocity of flow exists at a depth of 6/10 of the total depth, the length of wire or string connecting the two floats is at a depth of 6/10 of the total depth of flow. The velocity of flow is hence found by dividing the distance travelled by the float and the time taken to travel that distance.

b) Rod floats:

The rod float consists of a wooden rod or a hollow metallic rod weighed at the bottom so as to keep it vertical or inclined while travelling as shown in Figure (5-26).

The length of the rod should be so adjusted, by not touching the weeds at the bottom of the river, and its top will be above the water surface as shown.

A telescopic rod may be used to suit different depths. Since there is every possibility of the weed at of the river interfering with the rod float, hence a weed-free area is chosen.

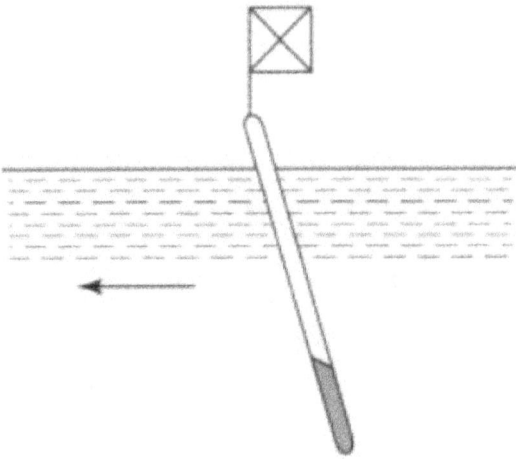

Fig(5-26) Rod floats

5.10.2 Pitot tube: A pitot tube is an instrument to determine the velocity at a required point in the flowing stream. In its simplest form, a pitot tube consists of a glass tube bent 90° as shown in Figure (5-27):

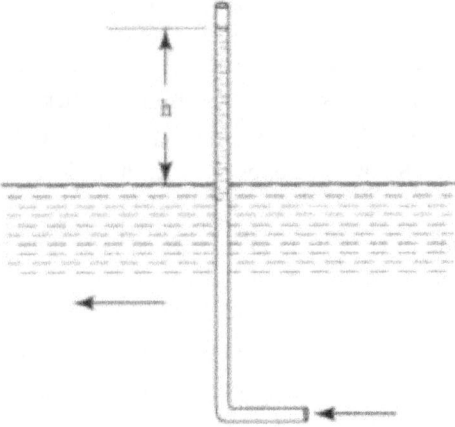

Fig (5-27) Pitot tube

The lower end of the tube faces the direction of flow. The water rises in the tube due to the exerted pressure by the flowing water. By measuring the rise of water in the tube, the velocity of the flowing water can be calculated using the following formula:

$$v = \sqrt{(2gh)}$$

Where,

v = velocity of water,

g = acceleration due to gravity, and

h = height of water in the tube above the surface

5.10.3 Chemical method of finding the discharge of a river: Another method of finding the discharge of a river or an irregular stream is by inserting uniformly a chemical solution (the solution will be such, which can easily mix up with the flowing water; generally common salt solution is used on account of its cheapness and easy mixing with water) into the flowing water. Great care must be taken in inserting the solution equally at several places over the cross-section of the stream. Then few samples of the water are taken, at lower sections, where the solution is evenly mixed with the flowing water:

Let,

Q = discharge of the stream in m^3/sec,

q = quantity of solution inserted in m^3/sec,

W = weight of salt per cubic meter of stream water at lower suction, and

w = weight of salt per cubic meter of solution inserted.

As the weight of salt, inserted per second, must be equal to the weight of salt over the lower section, therefore:

$$q\,w = Q\,W$$

Or,

$$Q = w\,/\,W\,q$$

5.10.4 Current meter: refer to sections: 5.1.2, 5.1.3, and 5.1.4

Exercise

1. A pipe of 5 cm diameter is suddenly enlarged to 10 cm diameter. Determine the loss of energy per kg of water, when the discharge is 20 litres/sec [**Ans.** 2.97 m]

diameter. If the discharge through the pipe is 1360 kg/min, determine:

(i) Loss of head at entrance,

(ii) Loss of head due to sudden enlargement, and

(iii) Loss of head due to sudden contraction.

[**Ans.** 21.4 cm; 24.2 cm; 13.4 cm]

3. An external cylindrical mouthpiece of 6 cm diameter is fitted to a vertical side of a tank containing water up to a height of 3 meters above the centre of the mouthpiece. Determine the discharge through the mouthpiece. [**Ans.** 12.85 litres/sec]

4. An internal mouthpiece of 6 cm diameter is discharging water under a head of 9 meters. Find the discharge in litres/sec, if the mouthpiece is: (i) running free, (ii) running full.
[**Ans.** 18.792 litres/sec; 26.57 litres/sec]

5. An external convergent-divergent mouthpiece is of 5 cm diameter. Determine the discharge, if the head of water over the mouthpiece is 11 meters. [**Ans.** 29 litres/sec]

6. A pipeline 3520 meters long and of 1 m diameter is supplying water to a hydroelectric plant under a net head of 535 meters. Find the velocity of water, if a nozzle of 20 cm diameter is fitted at the outlet end of the pipe. Take f = 0.006. [**Ans.** 48.97 m/sec]

7. A pipe 560 m long is discharging water 100 meters below the free surface of reservoir. Find the diameter of the nozzle, which will transmit maximum power. Assume coefficient of friction as 0.0054. [**Ans.** 10.17 cm]

8. A weir 30 meters long has a head of water 1 meter. Find the discharge over the weir, if the coefficient of discharge is 0.61. [**Ans.** 54.046 m^3/sec]

9. A weir 100 meters long is discharging water under a head of 1.25 meters. Using Bazin's formula, determine the discharge over the weir. [**Ans.** 252 m^3/sec]

10. A reservoir 50 meters long and 40 meters wide has been provided with a rectangular notch 1 meter wide at its base. Find the time taken to lower the water level of the reservoir from 15 cm to 10 cm. Take C_d= 0.63. [**Ans.** 2min 4.5sec]

10. A rectangular tank 25 meters long and 15 meters wide is fitted with a right angled V-notch on one of its sides. Determine the time required to lower the water level in the tank from 1 m to 48 cm. Take C_d= 0.62. [**Ans.** 3min 42sec]

11. A large vertical orifice 1.5 meters wide and 1 meter deep is discharging water from a tank. If the water level is 4 meters above the bottom edge of the orifice, find the discharge through the orifice. Take coefficient of discharge from the orifice as 0.62. [**Ans.** 7.71 m^3/sec]

12. A rectangular orifice 1.25 meters deep and 1 meter wide has difference of water levels on both sides of the orifice as 1.4 meters. Find the discharge through the orifice, if C_d= 0.6. [**Ans.** 381 m^3/sec]

13. Water issues horizontally from an orifice under a head of 16 cm. Find the coefficient of velocity of the jet, if the horizontal distance travelled by a point on a jet is 32 cm and the vertical distance is 17 cm. [**Ans.** 0.97]

14. A jet of water issues from a circular orifice of 2.5 cm diameter under a constant head of 1 meter. It falls 3.5 cm vertically down and strikes the ground at a distance of 35 cm from the centre of the vena contracta. Find the hydraulic coefficient, if the discharge through the jet is 1.35 litres/sec. [**Ans.** C_d= 0.625; C_v = 0.935; C_c = 0.668]

Check Your Knowledge

1) What are the various methods adopted for taking discharge measurement in a big river, channel and conduits.

2) Describe how float measurements are made to determine the flow in a river?

3) Describe with sketches the gauging of rivers by:

(i) float method.

(ii) current meter method, and

(iii) salt velocity method.

4) Describe the salt velocity method of gauging a river.

5) Write a short note on the use of the current meter method for the measurement of discharge rivers.

6) Draw a neat sketch of a current meter and explain how it is established.

6. Flow Through Pipes

6.1 Critical Velocity

It is a velocity at which the flow changes from the laminar flow to the turbulent flow. The critical velocity may be further classified into the following two types:

1) Lower critical velocity, and
2) Upper critical velocity.

6.1.1 Lower critical velocity: The change from laminar flow to turbulent flow occurs during a transition period between the two types of flows. The velocity at which the flow changes from laminar to the transition period is known as the 'Lower Critical Velocity'.

6.1.2 Upper critical velocity: A velocity at which the turbulent flow starts, or in other words, a velocity at which the flow enters from transition phase to turbulent flow is known as the 'Upper Critical Velocity'.

6.2 Reynolds Experiment of Viscous Flow

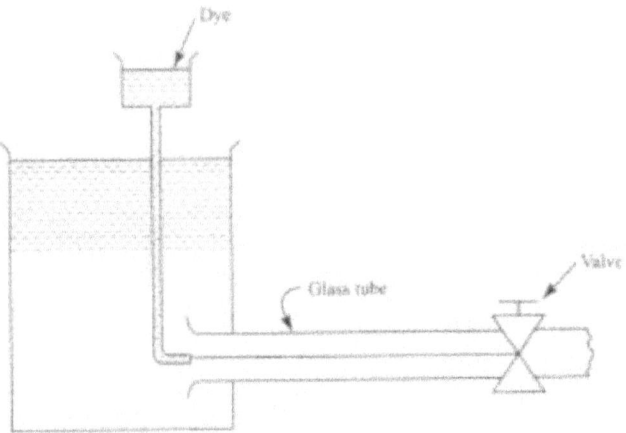

Fig (6-1) Reynolds apparatus

Reynolds apparatus consists of a tank, containing water and a small tank containing dye. A horizontal glass tube (1.5 meters long and 5 cm diameter) is fitted, through which water can flow. The water flow through the glass tube can be regulated by adjusting the regulating valve as shown in Figure (6-1).

The water in the tank is allowed to completely come to rest. The outlet valve of the glass tube is then slightly opened. A jet of dye, having the same specific gravity as that of water, is allowed to enter at the centre of the glass tube.

A fine thread of the dye is carried by the flowing water as shown in Figure (6-2) (a):

Fig (6-2) Three different shapes of flow inside the glass tube

The dye thread will steadily move, hardly can be seen to be in motion. Such a flow is known as 'laminar flow' or streamlines flow.

By increasing the velocity of water, through the glass tube, the dye thread will start becoming irregular and then breaking up as shown in Figure (6-2) (b). Such a velocity, at which the dye thread starts and becoming irregular, is known as 'lower critical velocity'. On further increasing the velocity of the water through the glass tube, the length of the dye thread in the glass will start decreasing and ultimately a stage at which the dye thread will disappear. Such a velocity, at which the whole dye thread is diffused, is known as an upper critical velocity. Beyond the upper critical velocity, the flow will be fully disturbed and such a flow is known as turbulent flow.

6.3 Loss of Head in Pipes

Water flowing in a pipe will experience some resistance to its motion whose effect is to reduce the velocity and ultimately the head of water available. Though there are many types of losses, the major loss will be due to frictional resistance of the pipe. The frictional resistance of a pipe depends on the roughness of the pipe. With more surface roughness of the pipe, greater will be the resistance. This friction is known as fluid friction and the resistance is known as frictional resistance.

Earlier experiments on fluid friction were conducted by Froude, who concluded that:

1) The frictional resistance varies approximately with the square of the velocity of the liquid, and
2) The frictional resistance varies with the nature of the surface.

Empirical formulae were derived for the loss of head due to friction, out of which the following are important from the subject point of view:

1) Darcy's formula for loss of head in pipes, and
2) Chezy's formula for loss of head in pipes.

6.3.1 Darcy's formula for loss of head in pipes:

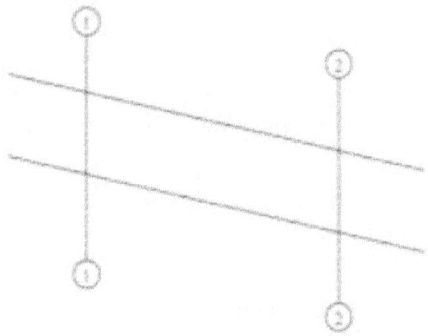

Fig (6-3) Uniform long pipe

Consider a uniform long pipe (> 100 x d) through which water is flowing at a uniform rate as shown in Figure (6-3):
Let,
L = length of pipe, and
d = diameter pf pipe,

Therefore, area of pipe, $A = \pi/4 \ (d)^2$

υ = velocity of water in the pipe,
f = friction resistance per unit area (of wetted surface) per unit velocity, and
h_f = loss of head due to friction.

At sections (1-1) and (2-2) of the pipe:

p_1 = intensity of pressure at section 1-1, and

p_2 = intensity of pressure at section 2-2.

p_1 would have been equal to p_2 at a state where no frictional resistance exist. Equating horizontal forces between sections (1-1) and (2-2) we get:

$$p_1 A = p_2 A + \text{Frictional Resistance}$$

Or,

$$\text{Frictional resistance} = p_1 A - p_2 A$$

(dividing both sides by w),

$$\text{Frictional resistance/w} = (p_1 A - p_2 A) / w$$

$$\text{Frictional resistance/Aw} = (p_1 A/w) - (p_2 A/w)$$

But,

$$(p_1 A/w) - (p_2 A/w) = \text{loss of pressure head due to friction} = h_f$$

Therefore,

h_f = friction resistance / Aw = friction resistance / $(\pi/4 \ (d)^2 \ w)$ ......(area, A = $\pi/4 \ (d)^2$)

From Froude's test,

Frictional resistance = frictional resistance per unit area at unit velocity x wetted area x (velocity)^2 = f x π d L x υ^2

Substituting the value of frictional resistance in the above equation:

$$h_f = f \pi d L \upsilon^2 / (\pi/4 \ (d)^2 \ w) = 4 f L \upsilon^2 / w d$$

Darcy recommended the following values of f:

(i) f = 0.005(1 + 1/12d) ….. for new and smooth pipes, and

(ii) f = 0.01 (1 + 1/12d) ….. for old and rough pipes.

Introducing a coefficient f', such that: f = f'w / 2g

Therefore,

$$h_f = 4/w d \ x \ f' \ w/2g \ x \ L \ \upsilon^2 = (4 f' L \upsilon^2) / 2gd \qquad\qquad (i)$$

As the discharge, Q = $\pi/4 \ (d)^2$ x υ

Or,

$$\upsilon = 4 Q / \pi d^2$$

Therefore,

$$\upsilon^2 = 16 Q^2/\pi^2 d^4$$

Substituting the value of υ^2 in equation (i) to get:

$$h_f = (4\ f'L/2gd) \times (16\ Q^2/\pi^2 d^4)$$
$$= f'L\ Q^2/3\ d^5 \qquad\qquad\qquad\qquad\qquad\qquad (ii)$$

Equations (i) and (ii) give us the major loss of head due to friction in pipes (equation (i) is suitable when the velocity of water in pipe is known, and equation (ii) is suitable when the rate of discharge of water is known).

In addition to the major loss of head due to friction, there are also minor losses of head such as:
(i) Loss of head at entrance = $0.5\ \upsilon^2/2g$, and
(ii) Loss of head due to velocity of water at outlet = $\upsilon^2/2g$.

Thus if all losses are considered, then:
$$h_f = 0.5\ \upsilon^2/2g + 4\ f'L\ \upsilon^2/2g\ d + \upsilon^2/2g$$

And by ignoring minor losses, we get:
$$h_f = 4\ f'L\ \upsilon^2/2g\ d$$

Chezy's formula for loss of head in pipes:

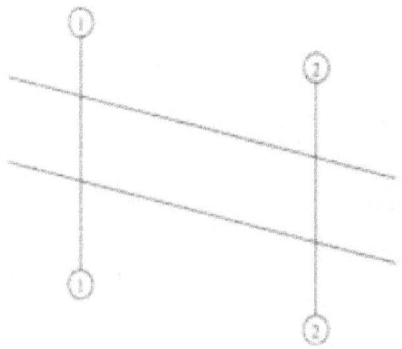

Fig (6-4) Uniform long pipe

Figure (6-4) shows a uniform long pipe, through which water is flowing at a uniform rate.
Let,
L = length of the pipe, and
d = diameter of the pipe.
Therefore,
Area of pipe, A = $\pi/4\ d^2$

And perimeter of the pipe, $P = \pi d$

υ = velocity of water in pipe,

f = frictional resistance, per unit area (of wetted surface) per unit velocity, and

h_f = loss of head due to friction.

Examining sections (1-1) and (2-2) 0f the pipe,

Let,

p_1 = intensity of pressure at section 1-1, and

p_2= intensity of pressure at sections 2-2

p_1 and p_2 would be equal if there was no frictional resistance. Taking into account the horizontal forces on water between sections (1-1) and (2-2) and equating the same:

$$p_1 A = p_2 A + \text{frictional resistance}$$

Or,

$$\text{Frictional resistance} = p_1 A - p_2 A$$

$$\text{Frictional resistance} / w = (p_1 A - p_2 A) / w \qquad \text{(dividing both sides by w)}$$

Or,

$$\text{Frictional resistance} / w = p_1/w - p_2/w$$

But,

$$p_1/w - p_2/w = \text{Loss of pressure head, due to friction} = h_f$$

Therefore,

$$h_f = \text{frictional resistance} / A\, w$$

From Froude's experiment:

Frictional resistance = frictional resistance per unit area at unit velocity x wetted area x (velocity)^2

Substituting the value of frictional resistance in the above equation, we get:

$$h_f = (f' \, x \, L \, \pi \, d \, L \, x \, \upsilon^2) / A\, w = (f' \, x \, P \, x \, L \, \upsilon^2) / A\, w = (f' \, L \, \upsilon^2 /w) \, x \, P/A \quad (\text{since } \pi d = P)$$

Replacing $A/P = m$, or $P/A = 1/m$ - where m is the hydraulic mean depth -:

$$h_f = (f' \, L \, \upsilon^2 /w) \, x \, L/m$$

Therefore,

$$\upsilon^2 = (h_f \, w \, m) / f' \, L = (w/f') \, x \, m \, x \, (h_f / L)$$

And,

$$\upsilon = \sqrt{w/f' \, x \, m \, x \, h_f/L}$$

Substituting $\sqrt{w/f'} = C$ (a constant known as Chezy's constant),

$h_f /L = i$ (where i is the loss of head per unit length)

Therefore,

$$\upsilon = C \sqrt{mi}$$

The hydraulic mean depth, for a circular pipe,

$$m = \text{Area/perimeter} = [\pi/4 \, (d)^2 / [\pi \, d] = d/4$$

6.4 Hydraulic Gradient and Total Energy Line

If pressure heads (i.e. p/w) of a liquid flowing in a pipe, is plotted as vertical ordinates on the centre line of the pipe, then the line joining the tops of such ordinates is known as 'hydraulic gradient' as shown in Figure (6-5):

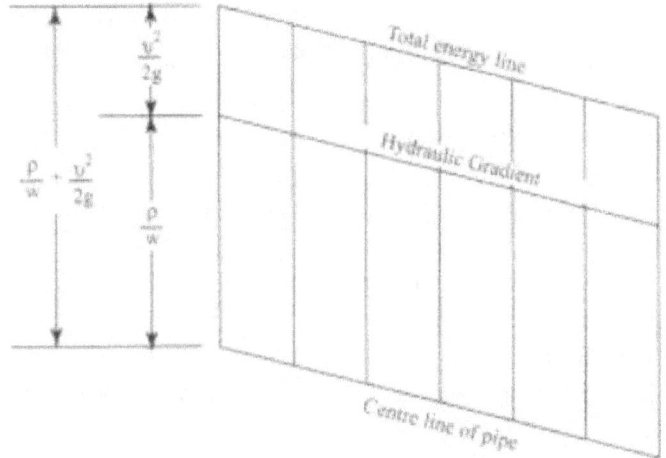

Fig (6-5) Hydraulic and total energy lines

If the sum of pressure head and velocity heads, (i.e. $p/w + \upsilon^2/2g$), of a liquid flowing in a pipe, is plotted as vertical ordinates on the centre line of the pipe, then the line joining the tops of such ordinates is known as 'total energy line'.

Or in other words, the total energy line lies over the hydraulic gradient, by an amount equal to the velocity head.

6.5 Boundary Layer Theory (or Prandtl's theory)

In earlier days scientists and engineers believed that when a body is held in the path of a moving fluid (or a body is moving through a fluid) will exert some shear force while flowing over its surface. This force causes rubbing action on the surface of the body.

This conception was found to be wrong by scientists of the earlier twentieth century. They argued that if there is a rubbing action of the fluid on the surface of the body, the body must show some fatigue and signs of decay due to continuous rubbing action. Prandtl was the first to publish his theory on boundary layer in 1904 (known as Boundary Layer Theory, or Prandtl's Theory). According to this theory, the liquid in the vicinity of the surface of the body may be divided into the following two portions:

(i) A very thin layer of the fluid, which is in immediate contact with the body; this layer of fluid behaves like a thin coating fixed to the boundary of the body with zero velocity. This thin layer of fluid is known as 'Boundary Layer'. The velocity of flow will rapidly increase in the normal direction to the surface. At the extreme end layer (i.e. outer edge of the boundary layer), the velocity of flow is approximately equal to the velocity of the fluid outside the boundary layer (more precisely, the velocity of flow is 0.99 times that outside the boundary layer).

The layer of flow in a pipe, in which there is velocity variation, is called a 'turbulent boundary layer'. Close to the pipe wall, however, there is a region in which the velocity approaches zero. Thus, although the whole layer may be classified as turbulent, the small laminar region close to the wall is frequently termed a 'Laminar Sub-layer'. This laminar sub-layer may have an appreciable thickness, or it may be so thin that the molecular viscous action is extremely small in comparison with the turbulent mixing in the central core of the pipe.

(ii) The remaining fluid, outside the boundary layer, has a high value of Reynold's number, because of the high velocity of flow.

6.6 Distribution of the Velocity of a Flowing Liquid over a Pipe Section

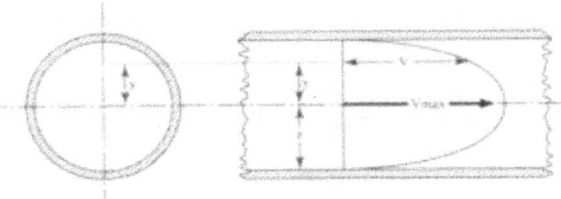

Fig (6-6) Velocity distribution of flowing liquid inside a uniform pipe diameter

Consider a flow in a pipe with uniform diameter. The velocity distribution diagram over the pipe section is as shown in Figure (6-6):

Let,

r   = radius of the pipe,

υ_{max} = maximum velocity of the liquid, and

υ = velocity of liquid at any other point at a radius y as shown in Figure (6-6)

The maximum velocity of a liquid is at its centre. Therefore, the maximum velocity is:

$$\upsilon_{max} = w\ H_L\ r^2/\ 4\ \mu\ L \tag{i}$$

Since the velocity at any section is:

$$\upsilon = (w\ H_L\ r^2/\ 4\ \mu\ L)\ (r^2 - y^2) \tag{ii}$$

Dividing equation (ii) by (i) we get:

$$\upsilon\ /\ \upsilon_{max} = (w\ H_L\ r^2/\ 4\ \mu\ L)\ (r^2 - y^2)\ /\ w\ H_L\ /\ 4\ \mu\ L)\ x\ r^2$$

$$= (r^2 - y^2)\ /\ r^2 = (1 - y^2/r^2)$$

$$\upsilon = \upsilon_{max}\ (1 - y^2/r^2)$$

6.7 Time of Flow from One Tank into Another through a Long Pipe

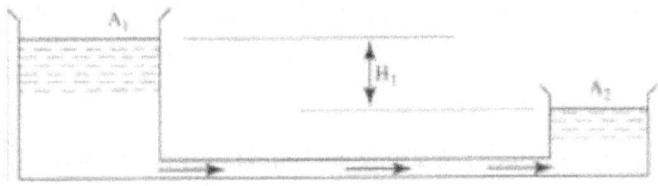

Fig (6-7) Two water tanks connected to a long pipe

Whenever two tanks, containing water, are connected together by a long pipe, the water will flow from the tank with a higher level to the tank with a lower level, irrespective of their areas.

In such a case the water level will drop in one tank, with a corresponding rise in the other. The head of water, causing flow in the pipe, will be the difference between the two water levels.

Consider two tanks connected by a long pipe as shown in Figure (6-7):
Let,
A_1 = area of the larger tank,
A_2 = area of the smaller tank,
L = pipe length,
d = pipe diameter,
a = pipe's area,
f = pipe's coefficient of friction,
H_1 = initial difference of water levels in the two tanks,
H_2 = final difference of water levels in the two tanks, and
T = time, in seconds, required to bring the difference of water levels from H_1 to H_2.

At some instant, let the difference of water levels between the two tanks be h. After a small interval of time dt, let the water level in the tank A_1 fall down by an amount equal to χ.

Therefore,
$$\text{Volume of water that passed from tank } A_1 = A_1 \, \chi \qquad \text{(i)}$$

And,
$$\text{The rise of water level in tank } A_2 = A_1 \, \chi \, / \, A_2$$

If the change of water levels in the two tanks be dh in time dt, then:

$$dh = \chi + (A_1 / A_2)\, \chi = \chi\, (1 + A_1/ A_2) = \chi[(A_1 + A_2) / A_2]$$

Or,

$$\chi = (- A_2 \, dh) / A_1 + A_2 \qquad \text{(ii)}$$

(minus sign indicate that the value h decreases as the discharge increases)

At an instant when the difference of water levels in the two tanks is h, then,

$$h = 4\, f\, L\, \upsilon^2/2g\, d, \text{ or } \quad \upsilon = \sqrt{(2\, g\, dh/4f\, L)}$$

Therefore,
The volume of water that has passed from tank A, in time dt = area of pipe x velocity x time
$$= a \sqrt{(2\, g\, dh/4f\, L)} \text{ x dt} \qquad \text{(iii)}$$

Equating (i) and (iii):

$$A_1 \, \chi = a_1 \sqrt{(2\, g\, dh/4f\, L)} \text{ x dt}$$

$$(-A_1 A_2 dh) / (A_1 + A_2) = a_1 \sqrt{(2\ g\ dh/4f\ L)} \times dt$$

(since, $\chi = (-A_2 dh) / A_1 + A_2$)

$$dt = (-A_1 A_2 dh) / a\ (A_1 + A_2) \sqrt{(2\ g\ dh/4f\ L)}$$
$$= \{- A_1 A_2 [\sqrt{(4\ f\ L/d)}(h)^{-1/2}\ dh]\} / \{a\ (A_1 + A_2)[\sqrt{2g}]\}$$

The total time (T) required bringing down the liquid levels from H_1 to H_2 may be found by integrating the above equation between the limits H_1 and H_2:

$$T = \int_{H_1}^{H_2} \{- A_1 A_2 [\sqrt{(4\ f\ L/d)}(h)^{-1/2}\ dh]\} / \{a\ (A_1 + A_2)[\sqrt{2g}]\}$$

The time T is also found as discussed below:

In T seconds, let the water level in tank A_1 fall by an amount equal to h.
Therefore,

      Volume of water that has passed from tank A_1 in T seconds = A_1 h   (i)

And,

      Volume of water that passed into Tank A_2 in T seconds = $A_2\ (H_1 - H_2 - h)$ (ii)

Since both volumes are equal, therefore:

$$A_1\ h = A_2\ (H_1 - H_2 - h) = A_2\ (H_1 - H_2) - A_2\ h$$

Or,

$$h\ (A_1 + A_2) = A_2\ (H_1 - H_2)$$

Or,

$$h = A_2\ (H_1 - H_2) / (A_1 + A_2)\qquad\qquad\text{(iii)}$$

Substituting the value of h in equation (i) to get:

The volume of water that passed from Tank A_1 in T seconds = $A_1 A_2 (H_1 - H_2)/(A_1 + A_2)$ (iv)

As the difference of water level H_1 is:

$$H_1 = 4\ f\ L\ \upsilon_1^2 / 2g\ d$$

Or,

$$\upsilon_1 = \sqrt{(2\ g\ d\ H_1)/(4\ f\ L)}$$

Similarly,

$$\upsilon_2 = \sqrt{(2\ g\ d\ H_2)/(4\ f\ L)}$$

Therefore,

$$\text{Average velocity} = [\sqrt{(2\ g\ d\ H_1)/(4\ f\ L)} + \sqrt{(2\ g\ d\ H_2)/(4\ f\ L)} / 2$$
$$= \tfrac{1}{2} [\sqrt{(2\ g\ d)/(4\ f\ L)}] (\sqrt{H_1} + \sqrt{H_2})\qquad\qquad\text{(v)}$$

Since, the time of flow, T = volume of water / Rate of flow

= volume of water / Area of pipe x average velocity

$= \{[A_1A_2(H_1 - H_2) / (A_1+ A_2)]\}/\{a \times \frac{1}{2} [\sqrt{(2\,g\,d)/(4\,f\,L)}\,] (\sqrt{H_1} +\sqrt{H_2}\,)\}$

$= 2A_1A_2\sqrt{4\,f\,L/d} / a\,(A_1+ A_2)\,\sqrt{2g}\,(\sqrt{H_1} - \sqrt{H_2}\,)$

$T = [-A_1A_2\sqrt{4\,f\,L/d}\,] / [a\,(A_1+ A_2)\,\sqrt{2g}\,\int_{H_1}^{H_2} h^{-1/2}\,dh$

$= [-A_1A_2\sqrt{4\,f\,L/d}\,] / [(1/2)\,a\,(A_1+ A_2)\,\sqrt{2g}\,[h^{-1/2}]_{H_1}^{H_2}$

$= [-2\,A_1A_2\sqrt{4\,f\,L/d}\,] / [a\,(A_1+ A_2)\,\sqrt{2g}\,] (\sqrt{H_1} - \sqrt{H_2}\,)$

Finally, ignoring the minus sign we get:

$T = [2\,A_1A_2\sqrt{4\,f\,L/d}\,] / [a\,(A_1+ A_2)\,\sqrt{2g}\,] (\sqrt{H_1} - \sqrt{H_2}\,)$

6.8 Discharge through a Compound Pipe (Pipes in Series)

Connecting pipes of different diameters, to form a pipeline, is known as 'compound pipe or pipes in series', as shown in Figure (6-8):

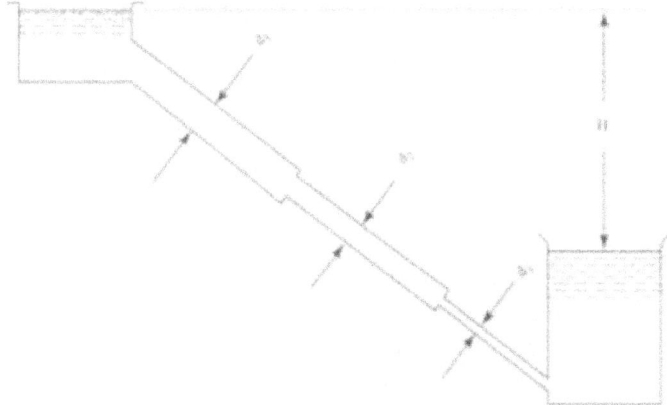

Fig (6-8) Pipeline with different diameters

Since these pipes are laid in series, therefore, the discharge will be continuous through these pipes.

Let,

Q = discharge through the pipe,

H = total loss of head,

h_{f1} = loss of head in pipe 1,

h_{f2} = loss of head in pipe 2,

h_{f3} = loss of head in pipe 3, and so on....

d_1 = diameter of pipe 1,

d_2 = diameter of pipe 2,
d_3 = diameter of pipe 3, and so on….
L_1 = length of pipe 1,
L_2 = length of pipe 2,
L_3 = length of pipe 3, and so on….
f_1 = coefficient of friction for pipe 1,
f_2 = coefficient of friction for pipe 2,
f_3 = coefficient of friction for pipe 3, and so on.

Neglecting the minor losses, except friction, hence the total loss of head is:

$$H = \text{(loss of head in pipe 1)} + \text{(loss of head in pipe 2)} + \text{(loss of head in pipe 3)} +….$$
$$= 4\, f_1\, L_1\, v_1^2/2g\, d_1 + 4\, f_2\, L_2\, v_2^2/2g\, d_2 + 4\, f_3\, L_3\, v_3^2/2g\, d_3 +….$$

If coefficient of friction is the same for all the pipes, then:

$$H = 4f/2g\, (L_1\, v_1^2/\, d_1 + L_2\, v_2^2/\, d_2 + L_3\, v_3^2/\, d_3 +….)$$

If the discharge through the pipeline is given, hence the total loss of head is:
$$H = f_1\, L_1\, Q^2/3d_1^5 + f_2\, L_2\, Q^2/3d_2^5 + f_3\, L_3\, Q^2/3d_3^5 + ….$$
$$H = Q^2/3\, (f_1\, L_1\, /\, d_1^5 + f_2\, L_2\, /\, d_2^5 + f_3\, L_3\, /3d_3^5 +….)$$

If the coefficient of friction is the same for all pipes, then:
$$H = f\, Q^2/3\, (L_1\, /\, d_1^5 + L_2\, /\, d_2^5 + L_3\, /3d_3^5 +….)$$

6.8.1 Discharge through pipes in parallel: In order to increase the discharge from one tank into the other, a new pipe is laid along the existing one. Such an arrangement is known as 'pipes in parallel' (Figure 6-9):

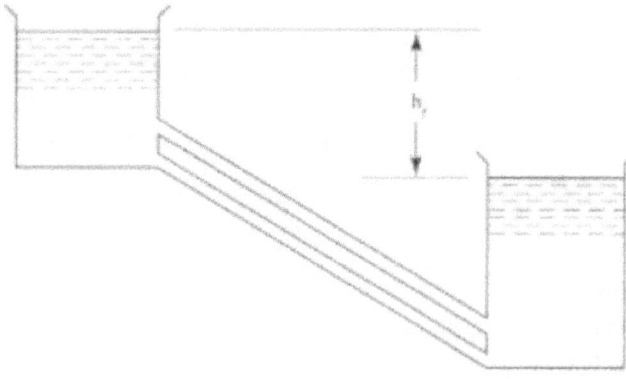

Fig (6-9) Pipes in parallel

As the pipes are laid in parallel, therefore, the loss of head will be the same while discharging water independently. The total discharge through all the pipes will be in this case the sum of the discharge in various pipes.

6.8.2 Discharge through branched pipes: Figure (6-10) shows a pipe lay partly with an existing pipe to increase the discharge in the lower tank:

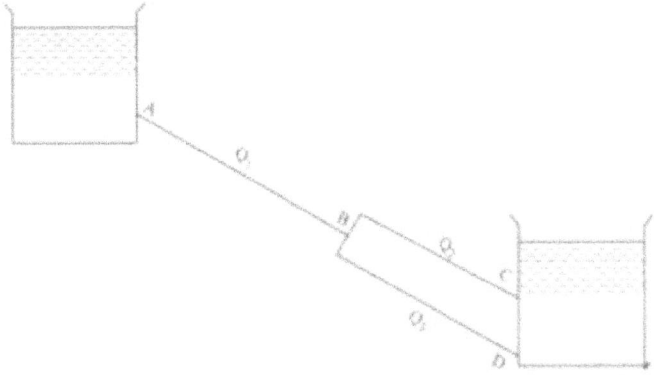

Fig (6-10) Pipe lay partly with existing pipe

In such a case, the discharge through the pipe 1 (i.e. Q_1) is equal to the sum of discharges through pipes 2 and 3 (i.e. $Q_2 + Q_3$).

Occasionally, an additional pipe (taking off from the main pipe) is feeding another tank, as shown in Figure (6-11):

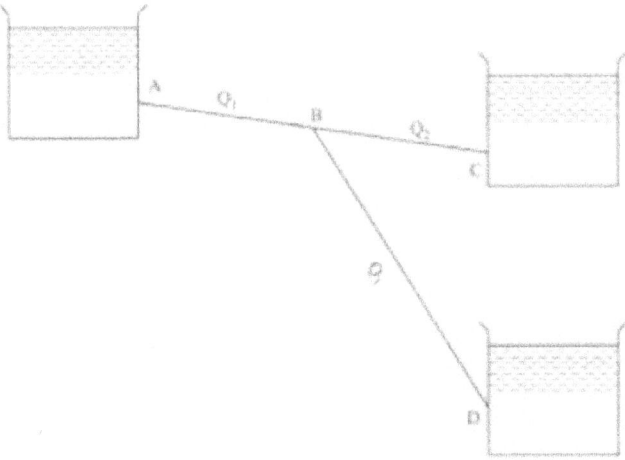

Fig (6-11) Additional pipe taking off from main pipe

In such a case, the discharge through the pipe 1 (i.e. Q_1) is also equal to the sum of discharges through pipes 2 and 3 (i.e. $Q_2 + Q_3$).

6.9 Transmission of Power through Pipes

Power is available whenever water is allowed to fall from a higher level to a lower one. Waterfalls are used to produce power by making the water flow through a pipe. This means that some head of water will be lost due to friction in the pipe.

Examining a high-level storage tank as in figure below:

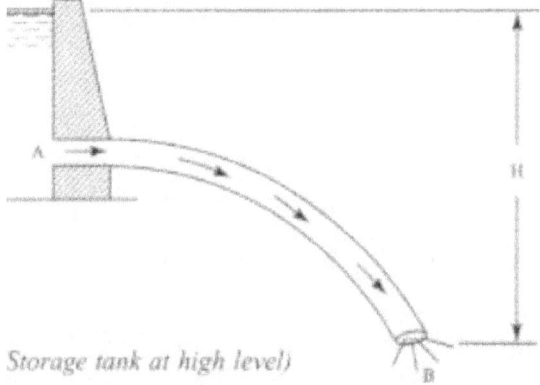

Storage tank at high level)

Let a pipe AB transmit water from A to a power house B as shown in figure above:
Let,
H = head of water at the power house in B meters,
L = length of pipe AB in meters,
v = velocity of water, in pipe, m/sec,
d = diameter of pipe Ab in meters,
a = cross-sectional area of the pipe in square meters = $\pi/4 \ (d)^2 \ m^2$,
h_f = loss of head in pipe AB, due to friction, and
f = coefficient of friction.

The weight of water flowing per second = w a v kg   (i)

Head of water available at B (neglecting minor losses) = H - h_f = H - $((4 f L v^2/2gd)$ m (ii)

As the efficiency of transmission is: $\eta = (H - h_f) / H$

And power, P = (weight of water flowing/sec x head of water) / 75
          = w a v [H - $(4 f L v^2/2gd)$] / 75   HP
          = w a [H v - $(4 f L v^2/2gd)$] / 75   HP   (iii)

(In the above equation, the power transmitted depends on the velocity of water, v)

Therefore, power transmitted will be maximum, when: $d(HP) / dv = 0$

Or, when the differential coefficient of the amount inside the bracket of equation (iii) is zero:

$$d[H\,v - (4\,f\,L\,v^2/2gd)] / dv = 0$$
$$H - 3(4\,f\,L\,v^2/2gd) = 0$$
$$H - 3 = 0 \qquad\qquad (\text{since, } 4\,f\,L\,v^2/2gd = h_f)$$

Or,

$$h_f = H/3$$

This means that power transmitted through the pipe is at a maximum power when the head lost due to friction in the pipe is equal to 1/3 of the total supply head.

6.10 Water Hammer

Water flowing in a pipe possesses some momentum on account of its motion. If the flowing water is suddenly brought to rest, by closing the valve, its momentum is stopped causing a very high pressure on the valve. This high pressure will be followed by a series of pressure vibration. The pressure vibration may cause noises in the pipe, known as 'knocking'.

Knocking in water pipes is often heard when a tap is turned off quickly. The sudden rise of pressure in a pipe is known as 'hammer blow or water hammer'. Sometimes the hammer blow is high enough to burst the pipe. In practice, it is important that valves of pipelines or penstock should always be closed gradually.

Examining a pipe through which the water is flowing with a uniform velocity:
Let,
L = length of the pipe
a = area of the pipe,
v = velocity of water in the pipe,
t = time, in seconds, in which the water is brought to rest by the closure of the valve.

Therefore,

$$\text{Mass of water in the pipe} = w\,a\,L / g \qquad\qquad \text{(i)}$$
$$\text{Rate of retardation of water} = \text{velocity} / \text{time} = v / t \qquad\qquad \text{(ii)}$$

Since,

$$\text{Force} = \text{mass x acceleration} = w\,a\,L/g \times v / t \qquad\qquad \text{(iii)}$$

Therefore,

Intensity of pressure rise in the valve, P = Force / Area

$$= (w\ a\ L/g \times v\ /\ t)\ /\ a = w\ L\ v\ /g\ t$$

6.11 Flow through Siphon Pipes

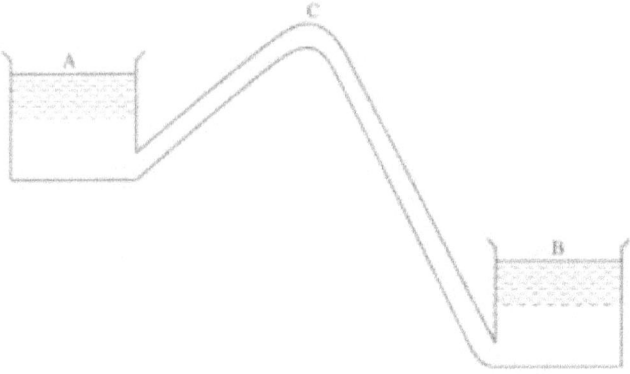

Fig (6-12) Summit of pipe above supply tank water level

If while laying a pipeline between two reservoirs, an obstacle in the form of a high ridge or high ground comes up, then it is not advisable to cut the ridge and lay the pipe; as the cost of laying the pipe will be more and the repairs will also be difficult. Thus the pipe is usually laid at the surface of the ridge. Sometimes it happens that the summit of the pipe is above the water level in the supply reservoir. Such a pipe is known as a siphon pipe as shown in Figure (6-12).

If the absolute pressure at C (least pressure in connecting pipe) is less than 2.5 meters of water, separation will occur, which will cause the flow to cease. The pipe should be laid in such a way that no section of the pipe will be more than 7.5 meters above the water level in tank A. (7.5 + 2.5 = 10 meters - suction limit)

Exercise

1) Water is being discharged from a tank through a pipe 165 meters long. The diameter for the first 25 meters is 5 cm and for the remaining 140 meters is 10 cm. Find the necessary height of the water surface, above the point of discharge, if the water is flowing with a velocity of 3 meters/sec. Take f = 0.0065 [**Ans.** 112.2 meters]

2) A compound pipeline 8 kilometres long is made up of pipes of 10 cm diameter for 1 kilometre, 20 cm diameter for 2 kilometres, 25 cm diameter for 1.5 kilometres and 30 cm diameter for 3.5 kilometres. It is required to replace the compound pipe by an equivalent pipe of the same length and for the same discharge. Find the diameter of the new pipe.

[**Ans.** 15 cm]

3) Two reservoirs having a difference in water level of 50 metres are connected by a pipeline 1500 meters long 60 cm diameter. Find the discharge through the pipe.

If a pipe 750 meters long and of 60 cm diameter is introduced parallel to the first and in the second half of the length, determine the increase in discharge.

[**Ans.** 685 litres/sec; 182 litres/sec]

4) Reservoir A (at an elevation of 220 meters above the datum) is supplying water to reservoir B (at an elevation of 130 metres above the datum) and C (at an elevation of 100 metres above the datum). The water is first flowing from reservoir A through a pipe 16 kilometres long and 30 cm diameter to a junction D. From D water flows to B through a pipe 10 kilometres long and 22.5 cm diameter and to C through a pipe 8 kilometres long and of 15 cm diameter. Find the discharge in the reservoirs B and C in litres/sec. Assume f = 0.01.

[**Ans.** 29.4 litres/sec; 15.2 litres/sec]

5) Find the head lost due to friction, in a pipe 500 meters long and 20 meters diameter, when the water is flowing with a velocity of 3 meters/sec. Take f = 0.01. [**Ans.** 45.8 m]

6) Water is flowing in a pipe 400 meters long and 15 cm diameter at the rate of 35.4 litres/sec. Determine the loss of head due to friction, if the coefficient of friction is 0.01.

[**Ans.** 21.8]

7) Water is flowing through a pipe 1500 meters long and 1 meter in diameter with a velocity of 1 meter/sec. Find the head lost due to friction by using:

a) Darcy's equation, with f = 0.005, and

b) Chazy's equation, with C = 64.

[**Ans.** 1.52 m; 1.42 m]

8) A horizontal pipe 600 meters long and of 30 cm in diameter connects two reservoirs, having a difference in water levels of 25 meters. Calculate the discharge through the pipe in litres/sec. Take f = 0.01. [**Ans.** 214.2 litres/sec]

8) Two reservoirs, having a difference of 45 meters in their levels, are connected by a siphon 2.7 kilometres long and 45 cm diameter. If the top of the siphon is two meters above the surface level of the upper reservoir, at a distance of 450 meters from the entrance, find the discharge through the siphon.

Take f = 0.005 and atmospheric pressure as 10.4 meters of water. [**Ans.** 375 litres/sec]

9) Two reservoirs, whose surface levels differ by 30 meters, are connected by a pipe of 60 cm diameter and 3 kilometres long. The pipeline crosses a ridge, whose summit is 9 metres above the level of water, and 300 meters distance from the higher reservoir. Find the minimum depth below the ridge, at which the pipe must be laid, if the absolute pressure in the pipe is not to fall below 3 meters of water. Also calculate the discharge through the pipe. Take f = 0.0075. [**Ans.** 4.9 m; 560 litres/sec]

10) Two reservoirs, whose surface levels differ by 100 meters, are connected by a pipe of 2 meters diameter and 10.000 meters long. The pipeline crosses a ridge, whose summit is 30 meters above the level of water in higher reservoir, and 500 meters distance from higher reservoir.

Find the minimum depth, below the ridge, at which the pipe should be laid, if the absolute pressure in the pipe is not to fall below 3 meters of water. Also find the discharge in litres/sec through the pipe considering all the losses. [**Ans.** 28.32 m; 11300 litres/sec]

7. Uniform Flow in Open Channels

7.1 Open Channel

An open channel is a passage through which the water flows under atmospheric pressure; or in other words when the free surface of the flowing water is in contact with the atmosphere, as in the case of a canal, a sewer or an aqueduct. A channel may be covered or open at the top.

The flow of water, in an open channel, is not due to any pressure, as in the case of pipe flow, but is due to slope of channel's head. Thus, during the construction of a channel, a uniform slope in the channel bed is provided to maintain the required water flow. Experiments have shown that the velocity of flow is different at different points in the cross-section of a channel, and all calculations are based on the mean velocity of flow.

7.1.1 Discharge through an open channel:

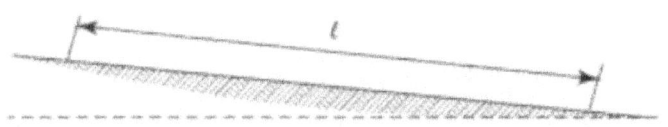

Fig (7-1) Open channel with uniform cross-section

An open channel of uniform cross-section and bed slope is shown in Figure (7-1):
Let,
L = length of the channel,
A = area of flow,
υ = velocity of water,
P = water perimeter of the cross-section,
f = frictional resistance per unit area at unit velocity, and
i = head's uniform slope.

Experiments have found that the total frictional resistance, in length L of the channel, follows a law:

$$\text{Frictional resistance} = f \text{ x contractual x (velocity)}^n$$
$$= f \text{ x P L x } \upsilon^n \qquad \text{(i)}$$

A practical value for 'n' is taken to be 2.
Therefore,

$$\text{Frictional resistance} = f \times P \ L \times \upsilon^2 \qquad\qquad \text{(ii)}$$

Since the water moves through a distance υ in one second,
Therefore,

Work done in overcoming the friction = frictional resistance x distance

$$= f \times P \ L \times \upsilon^2 \times \upsilon = \upsilon^3 \qquad\qquad \text{(iii)}$$

The weight of water in the channel in a length of L meters = w A L

Where,

w = sp. weight of water

This water will fall vertically down by a distance equal to (υ i) in one second,
Therefore,

Loss of potential energy = weight of water x height = w A L υ i

Or,

$$\upsilon^2 = (w \ A \ L) / f \ P$$

Or,

$$\upsilon = \sqrt{w/f} \times \sqrt{A/P \ i} = C \sqrt{m \ i}$$

Where,

$C = \sqrt{w/f}$ (known as Chezy's constant), and

m = A/P (known as hydraulic mean depth)

Therefore,

$$\text{Discharge } Q = A \times \upsilon = A \ C \sqrt{m \ i}$$

7.1.2 Values of Chezy's constant in the formula for discharge in open channel: The formula for flow in open channels was given by Chezy in 1775 as:

$$\upsilon = C \sqrt{m \ i}$$

Where, C is known as Chezy's constant and its value depends on the roughness of the inside surface of the channel. If the surface is smooth, there will be less frictional resistance to the motion of water, and as such the value of C will be more; which will result with more velocity or discharge and vice versa.

7.1.3 Manning's formula for discharge: Manning, after carrying out a series of experiments, deduced the following relation for the value of C in Chezy's formula for discharge:

$$C = 1/N \ m^{1/6}$$

Where, N is the Kutter's constant having values as shown in table below:

Values of N in the Kutter's formula

No.	Type of inside channel's surface	Value of N
1	Smooth cement plaster or planed wood	0.01
2	Concrete or unplanned wood	0.012
3	Poor brick or rubber stone	0.017
4	Earth or very good surface	0.02
5	Earth or ordinary surface	0.025
6	Earth of rough surface	0.03

Using Chezy's formula and Kutter's constant to determine the velocity:

$$v = C \sqrt{m\,i} = 1/N\ m^{1/6} \sqrt{m\,i}$$
$$= 1/N\ m^{1/6} \times m^{1/2} \times i^{1/2} = 1/N\ m^{2/3} \times i^{1/2}$$
$$= M\ m^{2/3} \times i^{1/2}$$

Where,

M = 1/N, known as Manning's constant.

Therefore,

$$\text{The discharge, } Q = \text{Area x velocity}$$
$$= A \times 1/N\ (m)^{2/3}\ (i)^{1/2}$$

7.2 Channels of Most Economical Cross-sections

A channel of most economical cross-section is that which gives maximum discharge for a given cross-sectional area and bed slop. In other words, it is a channel which involves lesser excavation for a designed amount of discharge.

A channel of most economical cross-section is also defined as a channel which has a maximum wetted perimeter; so that there is a minimum resistance to flow and thus resulting in a maximum discharge.

From the above, while deriving the condition for a channel of most economical cross-section, the cross-sectional area is assumed to be constant, and a relation between depth and breadth of the section is found out to give maximum discharge.

7.2.1 Condition of maximum velocity through a channel of circular section: Examining a channel of circular section, discharging water under the atmospheric pressure as shown in Figure (7-2):

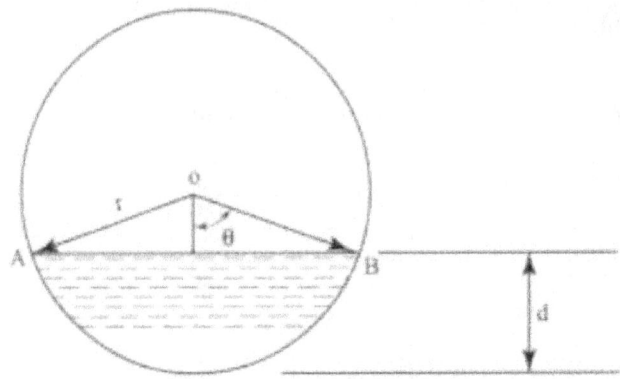

Fig (7-2) Channel with a circular section

Let,

r   = radius of the channel,

d   = depth of water in the channel, and

2Θ = total angle (in radius) subtended at the centre by the water surface AB

From the geometry of the figure, we find that the wetted perimeter of the channel:

$$P = 2\,r\,\Theta \qquad\qquad\qquad (i)$$

And area of the circular section, through which the water is flowing:

$$A = r^2\,\Theta - (r^2\,\sin2\Theta)/2 = r^2\,(\Theta - \sin2\Theta)/2 \qquad (ii)$$

The velocity of flow in an open channel is:

$$v = C\,\sqrt{m\,i} = C\,\sqrt{A/P\,i}$$

Keeping C and i constant, the velocity will be maximum when A/P is maximum:

$$d(A/P)\,/\,d\Theta = 0$$

Differentiating the above equation with respect to Θ and equating the same to zero:

$$(P\,dA/d\Theta - A\,dP/d\Theta)\,/\,P^2 = 0$$

Or,

$$P\,dA/d\Theta - A\,dP/d\Theta = 0$$

Substituting the values of P and A from equations (i) and (ii) to dA/dΘ and dP/dΘ by differentiating the equations (i) and (ii):

$$\{2r\,\Theta\,[r^2(1 - \cos 2\Theta)]\} - \{r^2\,[\Theta - (\sin 2\Theta/2) \times 2\,r]\} = 0$$
$$\{2r^3\,\Theta\,(1 - \cos 2\Theta)]\} - \{2r^3\,[\Theta - (\sin 2\Theta/2)]\} = 0$$

Dividing by $2r^3$:

$$[\Theta\,(1 - \cos 2\Theta)] - [\Theta - (\sin 2\Theta/2)] = 0$$
$$\Theta - \Theta \cos 2\Theta - \Theta + (\sin 2\Theta/2) = 0$$

Therefore,

$$\cos 2\Theta = \sin 2\Theta/2$$
$$\tan 2\Theta = 2\Theta$$

Solving the equation by trial and error, we get:

$$2\Theta = 257°\ 30'$$
$$\Theta\ = 128°\ 45'$$

Since, the depth of water, d = r - r cosΘ = r (1 - cosΘ)

$$= r\,(1 - \cos 128°\ 45') = r\,(1 + \cos 51°\ 15')$$
$$\text{(since, } \cos (180 - \Theta) = -\cos\Theta)$$
$$= r\,(1 + 0.62) = 1.62 \text{ radians} = 0.81$$

This means that the maximum velocity will take place when the depth of water is 0.81 times the diameter of the circular channel.

7.2.2 Conditions of maximum discharge through a channel of circular section:
Examining a channel of circular section, discharging water under the atmospheric pressure, as shown in Figure (7-3):

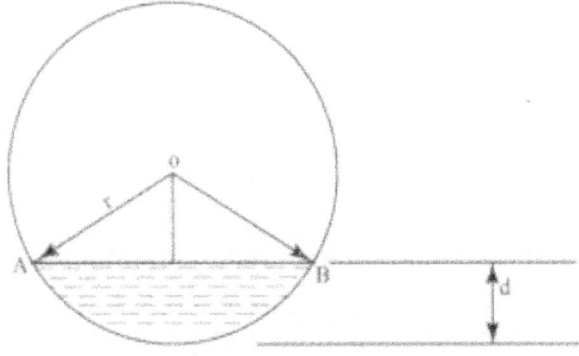

Fig (7-3) Channel with circular section

Let,

r   = radius of the channel,

d = depth of water in the channel, and

2Θ = total angle (in radius) subtended at the centre by the water surface AB.

From the geometry of the figure, we find that the wetted perimeter of the channel:

$$P = 2\,r\,\Theta \qquad\qquad\qquad (i)$$

And,

t = area of the circular section, through which the water is flowing.

$$A = r^2\,\Theta - (r^2\,\sin 2\Theta)/2 = r^2\,([\Theta - \sin 2\Theta/2)] \qquad (ii)$$

As the discharge through an open channel is:

$$Q = A\,C\,\sqrt{m\,i} = AC\,\sqrt{A/P\,i} = C\,\sqrt{A^3/P\,i}$$

Keeping C and i constant in the above equation, the discharge will be maximum when A^3/P is maximum.

Or,

$$d(A^3/P)/d\Theta = 0$$

Differentiating the above equation with respect to Θ and equating the same to zero:

$$(P \times 3A^2\,dA/d\Theta - A^3\,dP/d\Theta) / P^2 = 0$$

Multiplying by P^2 and dividing by A^2 :

$$3P\,dA/d\Theta - A\,dP/d\Theta = 0$$

Or,

$$3P\,dA/d\Theta = A\,dP/d\Theta$$

Substituting the values of P and A from equations (i) and (ii) to $dA/d\Theta$ and $dP/d\Theta$ by differentiating the equations (i) and (ii):

$$3 \times 2\,r\,\Theta \times r^2\,(1 - \cos 2\,\Theta) = r^2\,(\Theta - \sin 2\,\Theta/2) \times 2\,r$$
$$3\,\Theta \times 2\,r^3\,(1 - \cos 2\,\Theta) = 2\,r^3\,(\Theta - \sin 2\Theta/2)$$

Dividing by $2\,r^3$:

$$3\Theta \ (1 - \cos2\Theta) = \Theta - \sin2\Theta/2$$
$$3\Theta - 3 \ \Theta \cos2\Theta = \Theta - \sin2\Theta/2$$

Or,

$$2\Theta - 3 \ \Theta \cos2\Theta + \sin2\Theta/2 = 0$$

Solving this equation by trial and error:
$$\Theta = 154°$$

Since, the depth of water, $d = r - r \cos\Theta = r \ (1 - \cos\Theta)$
$$= r \ (1 - \cos 154° = r \ (1 + \cos 26°)$$
$$= 1.898 \text{ radians} = 0.949 \text{ diameter} \approx 0.95 \text{ diameter}$$

This means that the maximum discharge will take place when the depth of water is 0.95 times the diameter of the circular channel.

Check Your Knowledge

1) Explain the term open channel. Discuss the various types of open channels.

2) What do you understand by the term most economical section of a channel?

3) Find the best form of an open channel of rectangular section of given slope and area.

4) Drive an expression for condition of most economical section of a trapezoidal channel.

5) What is meant by channel of most economical section? Give three properties of such a section having a trapezoidal profile.

6) What are the conditions for maximum velocity and maximum discharge through a circular channel? Prove these conditions.

8. Non-uniform Flow in Open Channels

8.1 Introduction

In the previous chapter the uniform flow through open channels was discussed (i.e. a phenomenon of flow through open channels, in which the rate of flow, velocity of flow, depth of flow, area of flow, and slop of bed remains constant).

The change in any of the above conditions will cause the flow to be treated as non-uniform. An obstruction constructed across a channel of uniform width will also cause the flow to be non-uniform.

8.1.1 Specific energy: Specific energy is defined as the energy per unit weight (i.e. per kg) of the flowing liquid, with respect to the datum, passing through the bottom of the channel as shown in Figure (8-1):

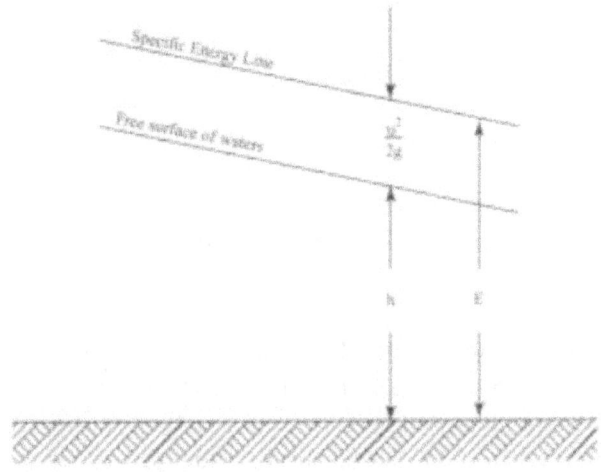

Fig (8-1) Specific energy line

Mathematically, specific energy is:

$$E = h + v^2/2g$$

Where,
h = depth of liquid flow, and
v = velocity of the liquid

8.1.2 Specific energy diagram: The specific energy of a flowing liquid is:

$$E = h + v^2/2g = E_s + E_k$$

Where,

E_s = h = static energy (also known as potential energy)

$E_k = v^2/2g$ = kinetic energy.

To plot specific energy diagram for a channel, such that depth of water along Y-Y axis and energy along X-X: we may conveniently do so by first drawing two independent curves, for static energy and kinetic energy, then adding their ordinates to get the required specific energy curve as shown in Figure (8-2):

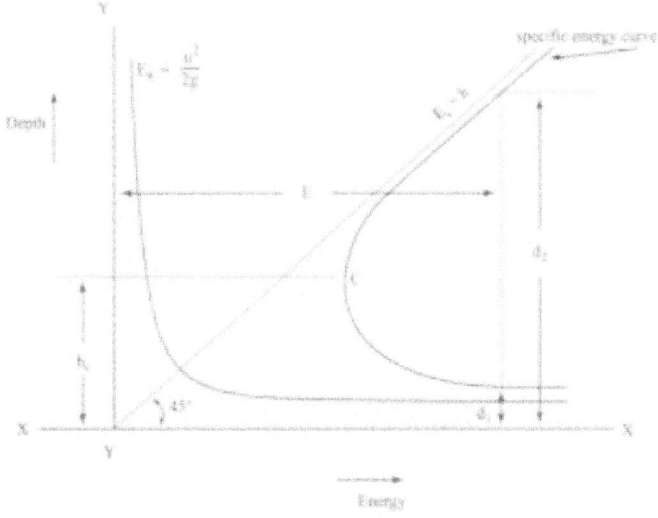

Fig (8-2) Specific energy diagram

Examination shows that the curve for static energy (i.e. E_s = h) will be a straight line, passing through the origin at 45° with the horizontal, and the curve for kinetic energy (i.e. $E_k = v^2/2g$) will be a parabola. By adding the values of these two curves, at all points, we get the specific energy curve.

8.2 Critical Depth

The specific energy diagram shows that the specific energy is a minimum at C. The depth of water in a channel, corresponding to the minimum specific energy (as at C in this case), is known as 'critical depth'. This depth is calculated by differentiating the specific energy equation and equating the same to zero:

$$dE/dh = 0$$
$$d(h + v^2/2g = 0 \qquad \text{(substituting E = h + } v^2/2g\text{)}$$
$$d/dh \ (h + q^2/h^2 \times 1/2g) = 0$$
$$(v = q/h, \text{ where } q = \text{discharge per unit width})$$

$$d/dh \ (h + q^2/h^2 \ x \ h^{-2}) = 0$$

Therefore,

$$[1 + q^2/h^2 \ x \ (-2) \ h^3] = 0$$
$$1 - q^2/gh^3 = 0$$
$$q^2/gh^3 = 1$$

Or,

$$v^2/gh = 1 \qquad\qquad (q^2/h^2 = v^2)$$

Or,

$$\text{Critical depth, } h_c = v_c^2/g \qquad\qquad\qquad (i)$$

The velocity of flow, at critical depth, is known as 'critical velocity' and is denoted by v_c.

Now for the minimum specific energy, substituting the value of h_c instead of h in the specific energy equation:

$$E = h + v^2/2g$$
$$E_{min} = h_c + v_c^2/g = h_c + h_c x \ g/2g$$
$$\text{(since, } h_c = v_c^2/g, \text{ therefore } v_c^2 = h_c x \ g)$$
$$= 3/2 \ h_c$$

Or,

$$\text{Static energy} = 2/3 \ E_{min} \qquad\qquad\qquad (ii)$$

Therefore,

$$\text{Kinetic energy} = E_{min} - 2/3 \ E_{min} = 1/3 \ E_{min} \qquad\qquad (iii)$$

We have seen in equation (i) that:

$$h_c = v_c^2/g = (q/h_c)^2/g \qquad\qquad (\text{since, } v = q/h)$$

Or,

$$h_c^3 = q^2/ \ g$$

Or,

$$h_c = (q^2/ \ g)^{1/3}$$

This is the required equation to find out the critical depth, when the unit discharge through the channel is given.

8.3 Hydraulic Jump

From the specific energy diagram it is proved that for a given specific energy E, there are two possible depths d_1 and d_2. Depth d_1 is less than the critical depth, and d_2 is greater than the critical depth.

When the depth is less than the critical depth, the flow is said to be a 'shooting flow'. When the depth is greater than the critical depth, the flow is said to be a 'streaming flow'. It is also found that the shooting flow is an unstable type of flow and does not continue on the downstream side. The flow transforms itself, to the streaming flow by increasing its depth.

The rise in the water level, which occurs during the transformation of the unstable shooting flow to the stable streaming flow, is called 'hydraulic jump' or 'standing wave'. At the hydraulic jump location most of the liquid energy is destroyed (mainly changed into heat energy). Hydraulic jump is known to be the best dissipater of water surplus energy. After the hydraulic jump, the water will flow with a greater depth and reduced velocity. The use of this phenomenon is applicable in hydraulic structures for irrigation purposes.

8.3.1 Depth of hydraulic jump: Examining two sections: upstream and downstream of a hydraulic jump, as shown in Figure (8-3):

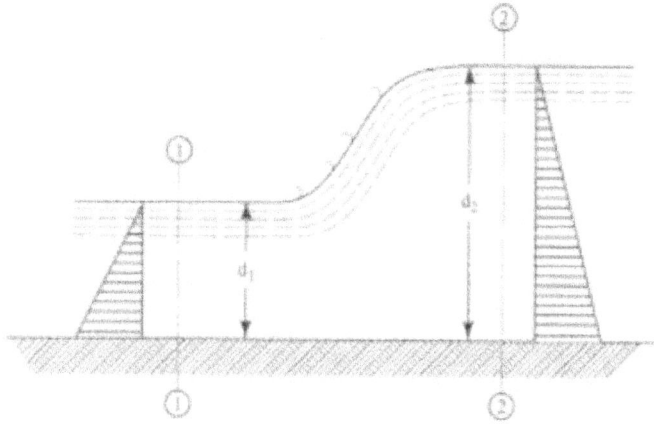

Fig (8-3) Hydraulic jump

Let,
1-1 = section on the upstream of the hydraulic jump,
2-2 = section on the downstream of the hydraulic jump,
d_1 = depth of flow at section 1-1,
d_2 = depth of flow at section 2-2,
v_1 = velocity of water at section 1-1,
v_2 = velocity of water at section 2-2, and
q = discharge per unit width.

$$q = Q/b$$

Where,
Q = total discharge
b = width of hydraulic jump

And,

$$q = d_1 \upsilon_1 = d_2 \upsilon_2$$

Considering the quantity of water between sections 1-1 and 2-2:

Force on section 1-1 [Force = w A x $\bar{x}$ = w (d_1 x L) x d_1/2 = (w d_1^2)/2] = (w d_1^2)/2
Similarly, force on section 2-2 = (w d_2^2)/2

The net force on the water column acting backward (because d_2 is greater than d_1)
$$= (w\ d_2^2)/2 - (w\ d_1^2)/2 = w/2\ (d_2^2 - d_1^2)$$

This net force is responsible for change of velocity from υ_1 to υ_2, and this force is also equal to = Mass of water flowing per second x change of velocity = w q/g (υ_1 - υ_2), where, q = discharge per unit width.

Or,

$$w/2\ (d_2^2 - d_1^2) = w\ q/g\ (\upsilon_1 - \upsilon_2)$$
$$(d_2^2 - d_1^2) = 2\ q/g\ (\upsilon_1 - \upsilon_2)$$
$$= w\ q/g\ (q/d_1 - q/d_2) = 2\ q^2/g\ [(d_1 - d_2)/\ d_1 d_2]$$

Or,

$$(d_2 + d_1)\ (d_2 - d_1) = 2\ q^2/g d_1 d_2\ (d_2 - d_1)$$

Or,

$$d_2 + d_1 = 2\ q^2/g d_1 d_2$$
$$d_2^2 + d_1 d_2 = 2\ q^2/g d_1 \qquad \text{(multiplying by } d_2)$$

Or,

$$d_2^2 + d_1 d_2 - 2\ q^2/g d_1 = 0$$

This is a quadratic equation in d_2
Therefore,

$$d_2 = (-d_1 \pm \{\sqrt{d_1^2 - [4\ x\ (2q^2/g\ d_1)]}\}\ /2$$
$$d_2 = d_1/2 + \sqrt{d_1^2/4 + 2q^2/g\ d_1} = -d_1/2 + \sqrt{d_1^2/4 + 2d_1 \upsilon_1^2/g}$$
(substituting q = $d_1 \upsilon_1$)

Depth of the hydraulic jump or height of the standing wave is $d_2 - d_1$.

8.4 Back Water Curve

Whenever an obstruction is placed across the width of the channel, the water surface on the upstream side of the obstruction no longer remains parallel to bed, and forms a curved surface. The amount by which the water level rises is known as afflux, and the curved surface (with concavity upwards) is called 'back water curve'. The distance between sections 1-1 and 2-2 (along the bed of the channel), shown in Figure (8-4), is known as length of the back water curve:

Where,

Section 1-1 = the section from where the backwater curve starts, and

Section 2-2 = the section where the backwater curve ends.

The afflux or back water curve occurs when:

(i) the loss of friction is less than the bed slope, and

(ii) decrease in the width of the channel.

8.4.1 Length of the backwater curve:

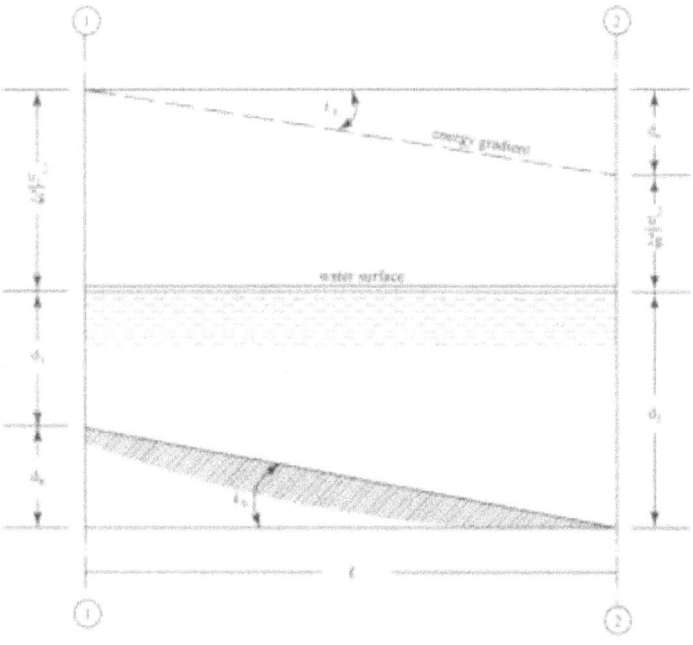

Fig (8-4) Channel with backwater curve

Examining a channel forming a backwater curve:

Let the length of the backwater curve be between the two sections 1-1 and 2-2 as shown in figure above.

Let,

v_1 = velocity of water at section 1-1,

d_1 = depth at flow at section 1-1

v_2 = velocity of water at section 2-2
d_2 = depth of flow at section 2-2
i_b = slope of the channel bed,
i_e = slope of energy gradient, and
L = length of the backwater curve.

Applying Bernoulli's equation at sections 1-1 and 2-2, we get:
$$d_b + d_1 + v_1^2/2g = d_2 + v_2^2/2g + d_e$$
$$(i_b \times L) + d_1 + v_1^2/2g = d_2 + v_2^2/2g + (i_e \times L)$$
Or,
$$L(i_b - i_e) = (d_2 + v_2^2/2g) - (d_1 + v_1^2/2g)$$
Or,
$$L = [(d_2 + v_2^2/2g) - (d_1 + v_1^2/2g)] / (i_b - i_e)$$
$$= (E_2 - E_1) / (i_b - i_e)$$

Where,
E_2 = specific energy at section 2-2, and
E_1 = specific energy at section 1-1

The energy gradient is determined by Manning's formula for an average velocity and average hydraulic mean depth.

Average velocity:
$$v_{av} = (v_1 + v_2) / 2$$
And,
Average hydraulic mean depth:
$$m_{av} = A_{av} / P_{av} \quad \text{or,} \quad (m_1 + m_2) / 2$$

Applying Manning's formula:
$$v_{av} = L/N \, (m_{av})^{2/3} (i_e)^{1/2}$$

8.5 Venturi-flume

In all hydraulic structures (i.e. bridge, regulator, etc...) constructed across an open channel, there are few openings left to allow water to pass. If the total width of all these openings is practically the same as that of the channel, such a structure is known as full width or inflamed structure. Generally, the total width of such a structure is kept much less than the width of the channel to have its effect on the economy during construction and to increase its utility. Such a structure, whose width is less than the width of the channel, is called 'flumed structure'.

A flumed structure (which is constructed across a channel by restricting its width) is used for the measurement of water quantity and is called 'venturi-flume'.

The following two types of venturi-flumes are important from the subject point of view:
(i) Non modular venturi-flume, and
(ii) Modular venturi-flume.

8.5.1 Non modular venturi-flume: It is a simple venturi-flume in which the width of the structure is reduced, as shown in Figure (8-5) (a,b):

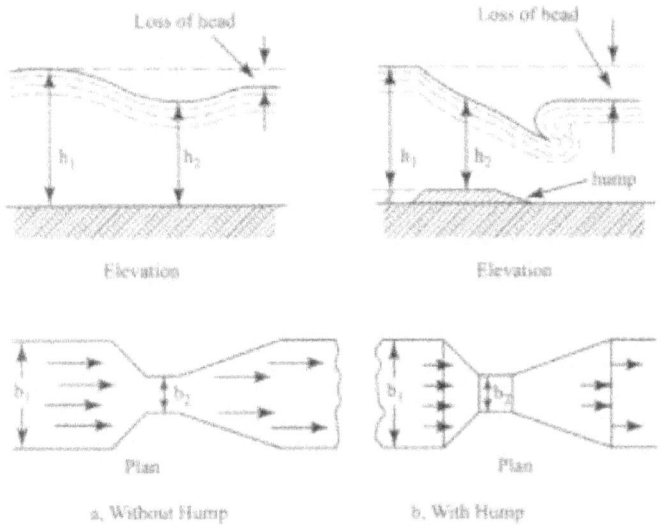

Fig (8-5) Venturi-flume: a) without a hump, b) with a hump

The restricted portion is called the throat. The theory of non-modular venturi-flume corresponds to that of the venturimeter (the only difference between a venturimeter and a venturi-flume is that in a venturimeter the flow is under pressure, whereas in the venturi-flume the flow is under gravity i.e. at atmospheric pressure).

In non-modular venturi-flume, there is loss of head as shown. A venturi-flume constructed in an open channel as shown in Figure (8-5) (a).

Let,
h_1 = depth of water at section 1,
b_1 = width of channel at section 1,
a_1 = area of flow at section 1, and
υ_1 = velocity of flow at section 1.

h_2, b_2, a_2, and v_2 = corresponding values at section 2.

Applying Bernoulli's equation at sections 1 and 2:

$$h_1 + v_1/2g = h_2 + v_2/2g \qquad \text{(taking } Z_1 = Z_2)$$
$$h_1 - h_2 = v_2^2/2g - v_1^2/2g \qquad\qquad\qquad \text{(i)}$$

Since the discharge at sections 1 and 2 continuous, therefore,

$$a_1 \, v_1 = a_2 \, v_2$$

Or,

$$v_1 = a_2 \, v_2/a_1$$

Therefore,

$$v_1^2 = (a_2^2 \, v_2^2) / a_1^2$$

Substituting the value of v_1^2 in equation (i):

$$h_1 - h_2 = v_2^2/2g - (a_2^2 \, v_2^2) /a_1^2 \text{ x } L/2g$$
$$= v_2^2/2g \, (1 + a_2^2/a_1^2 \,)$$
$$= v_2^2/2g \, (a_1^2 - a_2^2/a_1^2)$$

As $h_1 - h_2$ represents the difference in depth of water at sections 1 and 2, denoted by h:
Therefore,
$$h = v_2^2/2g \, (a_1^2 - a_2^2/a_1^2)$$

Or,

$$v^2 = (a_1^2/a_1^2 - a_2^2) \, 2gh$$

And,

$$v = (a_1/\sqrt{a_1^2 - a_2^2}) \, \sqrt{2gh}$$

The discharge through the venturi-flume:

$$Q = \text{coefficient of venturi-flume x } a_2 \, v_2$$
$$= C \, a_2 \, v_2 = (C \, a_1 a_2/\sqrt{a_1^2 - a_2^2}) \, \sqrt{2gh}$$

The following is to be noted:

1) The value of C (i.e. coefficient of venturi-flume) depends on the smoothness of the bed surface and sides as well as roundness of the corners. In general, the value of C lies between 0.95 and 1.0.

2) If the venturi-flume has a hump (a raised portion of the throat bed, which is generally of trapezoidal cross section) then the height of the hump should not exceed beyond its maximum value. Some authorities have fixed the maximum height of hump between 0.1 d_1 and 0.2 d_2, where d is the depth of water in the approach channel.

It is the author's opinion that the maximum height of hump should be 0.125 d_1 in its throat, and the depth of water at section L (i.e. h_1) is measured with reference to the bed of the throat i.e. top of the hump.

8.5.2 Modular venturi-flume: This is a particular type of venturi-flume in which the width of its throat, or depth of water at the throat, is decreased to such an extent that it is equal to the critical depth.

The velocity of flow through the throat, corresponding to the critical depth will also be critical. The existence of critical condition of flow in the throat will cause a standing wave on the downstream of the venturi-flume.

Such a venturi-flume which causes a standing wave on its downstream and has critical conditions in its throat is known as modular venturi-flume, or standing wave venturi-flume as shown in Figure (8-6) (a,b):

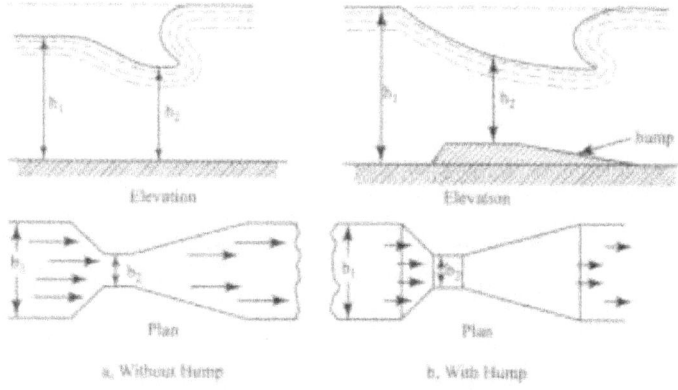

Fig (8-6) Standing wave venturi-flume: a) without a hump, b) with a hump

In a modular venturi-flume there is always some rise of water on its downstream as shown. Examining a modular venturi-flume constructed in an open channel:

Let,
h_1 = depth of water at section 1,
b_1 = width of channel at section 1,
a_1 = area of flow at section 1, and
v_1 = velocity of flow at section 1.

h_2, b_2, a_2, and v_2 = corresponding values at section 2.

Therefore,

$$\text{Specific energy at section 1, } E_{S1} = v_1^2/2g$$

Similarly,

$$E_{S1} = h_2 + v_2^2/2g$$

Neglecting the loss of energy in the gradually converging portion of the flume, we get:

$$E_{S1} = E_{S2}$$

Therefore,

$$E_{S1} = h_2 + v_2^2/2g \qquad\qquad\qquad (i)$$

Or,

$$v_1^2/2g = E_{S1} - h_2$$

Therefore,

$$v_2^2 = 2g\,(E_{S1} - h_2)$$

Therefore,

$$v_2 = \sqrt{2g(E_{S1} - h_2)} \qquad\qquad\qquad (ii)$$

The theoretical discharge through the venturi-flume:

$$Q_{th} = a_2\,v_2 = (b_2\,h_2) \times \sqrt{2g(E_{S1} - h_2)}$$
$$= b_2\,h_2\,\sqrt{2g} \times \sqrt{(E_{S1} - h_2)} \qquad\qquad (iii)$$
$$= b_2\sqrt{2g} \times \sqrt{E_{S1}h_2^2 - h_2^3}$$

Since, the depth of water in the throat (i.e. h_2) of this venturi-flume is critical, hence the discharge through the venturi-flume must be maximum.

With maximum discharge:

$$dq/dh_2 = 0$$
$$d/dh_2[b\,\sqrt{2g} \times \sqrt{E_{S1}h_2^2 - h_2^3} = 0$$
$$d/dh_2[\sqrt{E_{S1}h_2^2 - h_2^3} = 0 \qquad\qquad \text{(since, } b\,\sqrt{2g} \text{ is constant)}$$
$$E_{S1} \times 2h_2 - 3h_2 = 0$$
$$h_2(2E_{S1} - 3h_2) = 0$$

Or,

$$2\,E_{S1} - 3h_2 = 0 \qquad\qquad \text{(since, } h_2 \text{ cannot be zero)}$$
$$h_2 = 2/3\,E_{S1} \qquad\qquad\qquad (iv)$$

From equation (i) we have:

$$E_{S1} = h_2 + v_2^2/2g \quad \text{or,} \quad v_2^2/2g = E_{S1} - h_2$$

Substituting the value of h_2 from equation (iv):
$$v_2^2/2g = 1/3 \ E_{S1}$$

From equation (iii) we have:
$$Q_{th} = b_2 \ h_2 \ \sqrt{2g} \ x \ \sqrt{(E_{S1} - h_2)}$$
$$= b_2 \ x \ 2/3 \ E_{S1}\sqrt{2g} \ x \ \sqrt{(E_{S1} - 2/3 \ E_{S1})}$$
$$= b_2 \ x \ 2/3 \ E_{S1}\sqrt{2g} \ x \ (1/\sqrt{3}) \ (E_{S1})^{1/2}$$
$$= (2/3\sqrt{3}) \ (\sqrt{(2g)}b_2(E_{S1})^{3/2} = 1.7 \ b_2 \ (E_{S1})^{3/2}$$

Therefore,
$$\text{Actual } Q = C \ x \ 1.7 \ b_2 \ (E_{S1})^{3/2}$$

Where,
$$C = \text{coefficient of venturi-flume.}$$

The following to be noted:

1) The value of C depends on the smoothness of the bed surface and sides as well roundness of the corners. In general the value of C lies between 0.95 and 1.0.

2) If the venturi-flume has a hump in its throat, the depth of water at section 1 (i.e. h_1) is measured with reference to the head of the throat i.e. top of the hump.

Check Your Knowledge

1) Define the term 'specific energy'.
2) What are specific energy diagrams? How are they useful in the phenomenon of flow in an open channel?
3) Define the term critical depth as applied to the flow in an open channel.
4) What is a hydraulic jump? Explain clearly how it is found.
5) What are the conditions favourable for the formation of a hydraulic jump in an open channel?
6) Drive an expression for obtaining the depth of flow in the channel on the downstream side of the hydraulic jump: $d_2 = - d_1/2 + \sqrt{[d_1^2/4] + [2 \ v_1^2 \ d_1)g]}$
7) Why is it desirable to have a hydraulic jump at the top of a spillway? How can this be attained?
8) What do you understand by the term 'Back-water curve'? Drive an equation for finding out the length of the back-water curve.
9) What is the use of venturi-flume? Distinguish between modular and non-modular venturi-flumes.

9. Motion of Bodies in Fluids

9.1 Introduction

When a solid body is held in the path of a moving fluid and is completely immersed in it, the body will be subjected to some pressure or force. Conversely, if a body is moved through a fluid at rest it offers some resistance to the moving body, or the body has to exert some force to maintain its steady movement. Hence, when a submarine moves through the water or an aeroplane flies through the atmosphere, its engine must supply a sufficient force to balance the present resistance.

9.2 Pressure on an Immersed Body in a Moving Liquid

Whenever a body is immersed in a moving liquid it exerts some pressure on the immersed body. If a plate is immersed in a liquid, parallel to the flow, it will be subjected to a pressure less than if the same plate is held immersed perpendicular to the flow.

But if the same plate is immersed at some angle with the direction of flow, the streamlines of the liquid get deflected as shown in Figure (9-1):

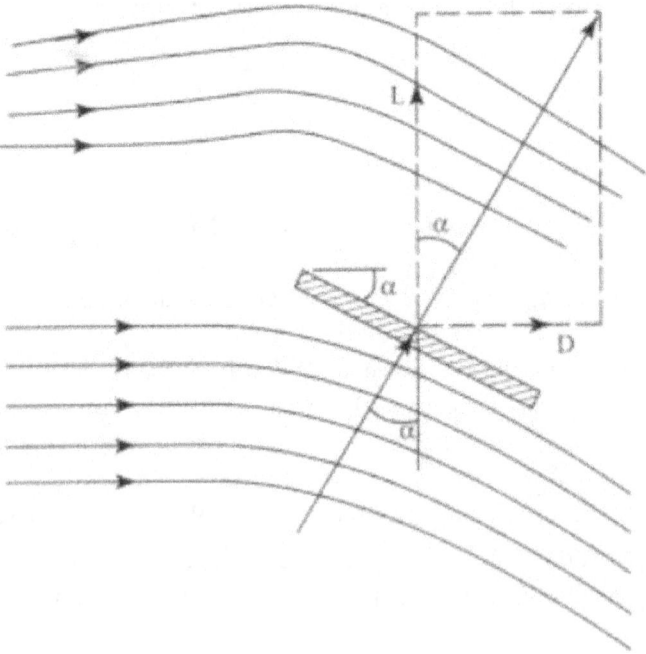

Fig (9-1) Plate immersed at some angle to the direction of flow

As per Pascal's law, the pressure exerted by the moving liquid will be at right angles to the plate. In such a case, the following two types of pressure are important from the subject point of view:

(i) Drag, and

(ii) Lift.

9.2.1 Drag: Whenever a plate is held immersed at some angle with the direction of flow it is subjected to some pressure. As this pressure acts at right angles to the plate, therefore it will possess the following components:

(i) in the direction of flow, and

(ii) at right angles to the direction of flow.

The component of this pressure in the direction of flow is known as 'drag'.

Consider a plate held immersed in the flow of a liquid as shown in Figure (9-1):

Let,

a = area of plate in m^2,

w = specific weight of the flowing liquid,

V = component of the velocity of liquid at right angles to the plate in m/s, and

α = angle at which the plate is inclined with the vertical.

Therefore,

$$\text{Mass of liquid exerting pressure on the plate} = w\,a\,V/g \text{ kg, and}$$

The velocity of liquid, after exerting pressure at right angle to the plate will be reduced to zero.

Therefore,

Pressure of liquid on plate, P = mass of liquid flowing/sec x change of velocity

$$= w\,a\,V/g\,(V - 0)$$

$$= w\,a\,V^2/g \text{ kg}$$

The component of this pressure in the direction of flow, i.e. drag (If the liquid is at rest and the immersed plate is moving with a velocity V in the liquid, then the drag may be defined as the horizontal resistance offered by the liquid to the moving plate):

$$D = (w\,a\,V^2/g)\sin\alpha = K_D \times w\,a\,V^2/g$$

Where, K_D is a coefficient, known as drag coefficient (sometimes it is called coefficient of drag). Its value depends upon the type of plate and the angle of inclination of the plate, and is determined experimentally.

9.2.2 Lift: Whenever a plate is held immersed at some angle with the direction of flow of the liquid, it is subjected to some pressure, which acts at right angles to the plate. The component of this pressure at right angle to the direction of flow of the liquid is known as lift.
Since,

$$\text{pressure } P = w \, a \, V^2/g$$

The component of this pressure at right angles to the flow, i.e. lift, [if the liquid is at rest and the immersed plate is moving with a velocity V in the liquid, then the lift may be defined as the vertical resistance offered by the liquid to the moving plate. This creation of lift, on the bodies, is used for propelling ships and for supporting the weight of the flying aeroplane]:

$$L = (w \, a \, V^2/g) \cos\alpha = K_L \, x \, w \, a \, V^2/g$$

Where, K_L is coefficient known as lift coefficient (or coefficient of lift). Its value depends on the type of plate and its angle of inclination, which is determined experimentally. The resultant force on the body is:

$$R = \sqrt{D^2 + L^2}$$

9.3 Boundary Layer Separation

When a body is held immersed in a flowing liquid, a thin layer of the liquid will behave as if it is fixed to the boundary of the body. But if the immersed body is a curved or angular one, the boundary layer does not stick to the whole surface of the body. The boundary layer leaves the surface and separates. This phenomenon is known as 'boundary layer separation'. The point where the boundary layer separates from the surface of the body is known as 'point of separation'.

9.4 Prandtl's Experiment of Boundary Layer Separation

Post his publication of boundary layer theory, Prandtl conducted series of experiments on boundary layer separation. A cylinder was held in flowing liquid sprinkled with minute particles of aluminium (to photograph streamlines) on the surface of the liquid. This allowed the liquid to flow around the cylinder and the boundary layer adhered to the surface of the cylinder throughout. The liquid streamlines flu around the cylinder as shown in Figure (9-2 a):

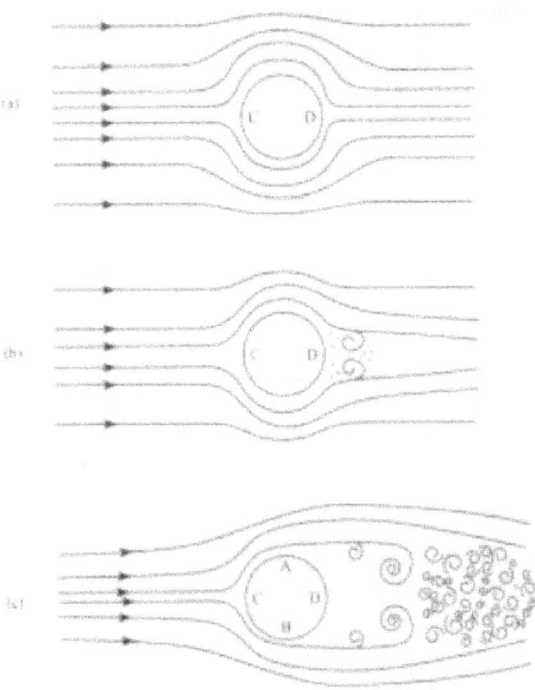

Fig (9-2) Streamlines around the cylinder

By gradually increasing the velocity of liquid, at a certain velocity the streamlines at D became irregular. By increasing further the velocity of liquid, the boundary layer separate from the surface on both sides of D and vortices (a flow, in which the liquid flows continuously round a curved path and about a fixed axis of rotation, is called vortex flow) formed as shown in Figure (9-2 b). Further increase of velocity will lead to an earlier separation of boundary layer forming more pronounced vortices. In the final stages of the boundary layer separation takes place at points A and B as shown in Figure (9-2 c).

9.5 Magnus Effect in a Moving Liquid

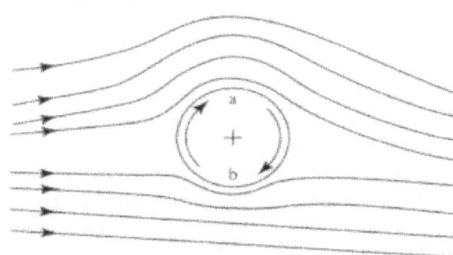

Fig (9-3) Streamlines flowing over a rotating cylinder

A liquid having streamline flow from left to right; let a cylinder be rotated about its longitudinal axis in the path of the streamlines. The rotating motion of the cylinder deviate the streamlines as shown in Figure (9-3). The phenomenon of deviating of the streamlines by the rotating cylinder is known as 'Magnus Effect'.

The velocity of liquid at 'a' has increased, because of the movement of the cylinder, which exerts a viscous drag on the liquid and thus increases its velocity. The velocity of liquid at 'b' is reduced because of the viscous drag on the liquid by the moving cylinder in the opposite direction.

9.6 Prevention of Boundary Layer Separation by Providing Slots near the Leading Edge

The boundary layer separation may be prevented by providing slots on the wings of an aeroplane (or any other similar body) near the leading edge.

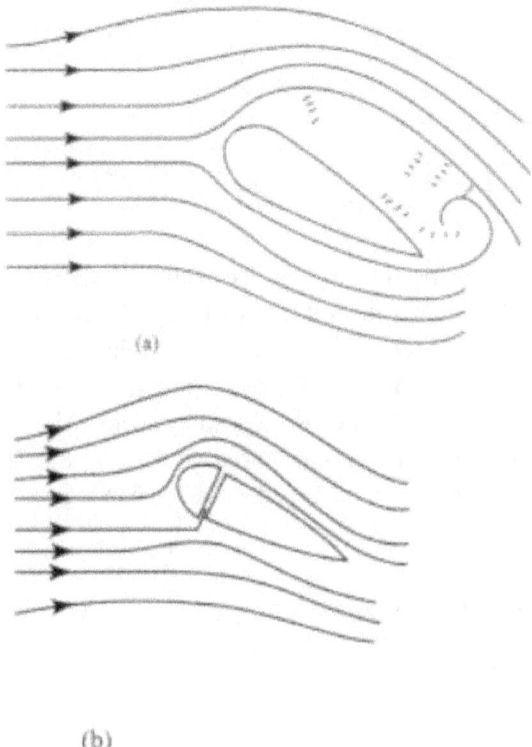

(a)

(b)

Fig (9-4) Flow over an aeroplane wing - aerofoil

Figure (9-4)(a) shows the separation of boundary layer. Figure (9-4)(b) shows a slot made near the leading edge, while the remaining conditions remain unchanged. Because of the slot, the air bands pass through it and adhere on the upper surface of the wing for the whole length.

9.6.1 Prevention of boundary layer separation by providing rotating cylinder as the leading edge: The boundary layer separation may also be prevented by providing rotating cylinder as the leading edge of the aeroplane wings (or any other similar body).

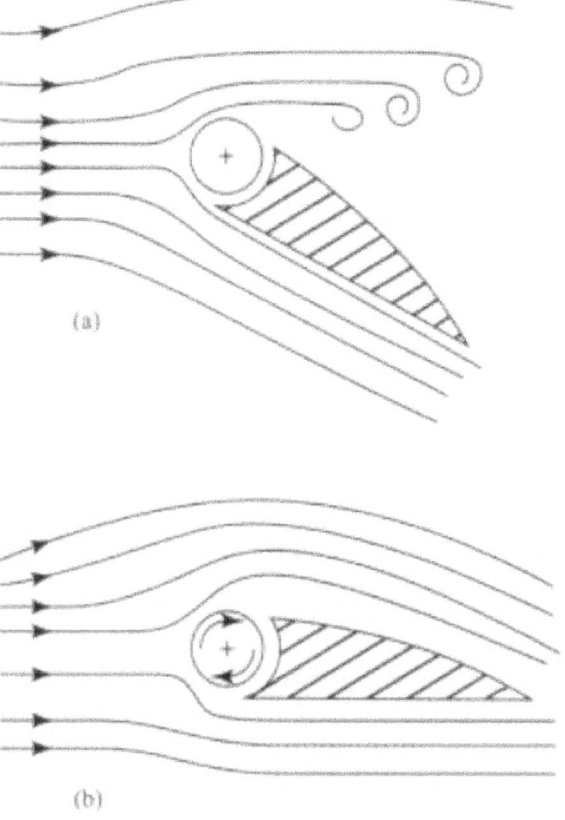

Fig (9-5) Rotating cylinder at leading edge of aerofoil

Figure (9-5)(a) shows the separation of boundary layer, when the cylinder is at rest. Figure (9-5)(b) shows the rotating cylinder, while the remaining conditions are unchanged. Because of the rotating cylinder, the Magnus effect takes place on the air bands, which adhere on the upper surface of the wing for the whole of its length.

Check Your Knowledge

1) Write short notes on lift and drag.

2) Explain clearly, with sketches, the growth of boundary layer theory in a turbulent flow past a flat plate placed parallel to the flow.

3) What do you understand by the term thickness of boundary layer? Derive an equation for the thickness of boundary layer in a laminar flow.

4) Explain clearly the phenomenon of boundary layer separation.

5) What is meant by Magnus effect?

6) Discuss the methods adopted to prevent the boundary layer separation.

10. Impact of Jets

10.1 Introduction

Whenever a jet of liquid impinges (i.e. strikes) on a fixed plate, the plate experience some force. This force is equal to the rate of change of momentum of the jet. It was observed that if the plate is not fixed, it will start moving in the direction of the jet.

10.1.1 Force of jet impinging normally on a fixed plate: Consider a jet of water impinging normally on a fixed plate as shown in Figure (10-1):

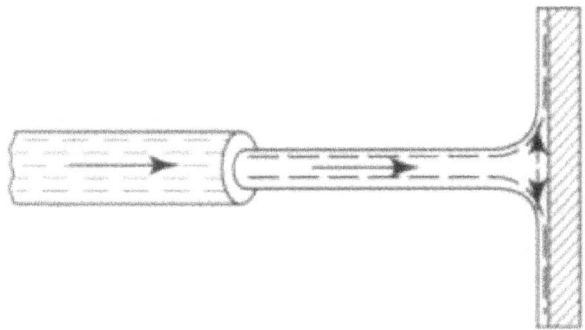

Fig (10-1) Jet of water impinging normally on a fixed plate

Let,
V = velocity of the jet in m/s, and
a = cross-sectional area of the jet in m^2,
Therefore,

$$\text{Mass of water flowing per second} = w\, a\, V/g \qquad (i)$$

As the velocity of jet, in its original direction, is reduced to zero after impact, therefor,

Force exerted by the jet on the plane, F = mass of water flowing per sec x change of velocity
$$= w\, a\, V/g \times (V - 0)\, kg = w\, a\, V^2/g \ \ kg$$

10.1.2 Force of jet impinging on an inclined plate: Consider a jet impinging on an inclined fixed plate as shown in Figure (10-2):

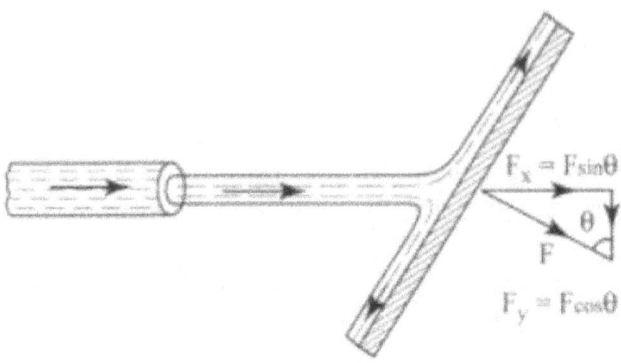

Fig (10-2) Jet impinging on an inclined plate

Let,

V = velocity of jet in m/s,

a = cross-sectional area of the jet in m^2, and

Θ = angle at which the first plate is inclined with the jet.

As the force exerted by the jet in its original direction,

= mass of water flowing per second x change of velocity

= w a V/g x (V - 0) = w a V^2/g   kg

The force exerted by the jet in a direction normal (i.e. perpendicular) to the plate:

$$F = (w\,a\,V^2\sin\Theta) / g$$

And the force exerted by the jet in the direction of flow,

$$F_x = F\sin\Theta = [(w\,a\,V^2\sin\Theta) \times \sin\Theta] / g$$
$$= (w\,a\,V^2\sin^2\Theta) / g$$

Similarly, the force exerted by the jet in a direction normal to the flow is:

$$F_y = F\cos\Theta = [(w\,a\,V^2\cos\Theta) \times \cos\Theta] / g$$
$$= (2\,w\,a\,V^2\sin^2\Theta\,\cos\Theta) / 2g$$
$$= (w\,a\,V^2\,\sin2\Theta) / 2g \qquad \text{(since, } \sin2\Theta = 2\sin\Theta\,\cos\Theta)$$

10.1.3 Force of jet impinging on a moving plate: Consider a jet of water impinging normally on a plate. As a result of the impact of the jet, let the plate move in the direction of the jet as shown in Figure (10-3):

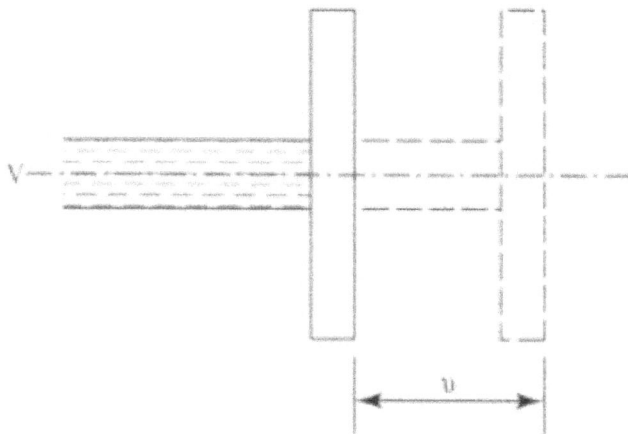

Fig (10-3) Moving plate as a result of impinging water

Let,

V = velocity of the jet in m/s,

a = cross sectional area of the jet m^2, and

υ = velocity of the plate as a result of the impact of the jet in m/s.

Therefore,

Relative velocity of the jet with respect to the plate = (V - υ) m/s

For analysis purposes, it will be assumed that the plate is fixed and the jet is moving with a velocity (V - υ) m/s.

Since the force exerted by the jet is:

F = Mass of water flowing per second x change of velocity

= [w a(V- υ)] / g x [(V - υ) - 0]   kg

= [w a(V - υ)^2] / g kg

And the work done by the jet = Force x distance

= [w a(V- υ) υ] / g   kg-m

10.1.4 Force of jet impinging on a series of vanes: Let a jet of water impinge on a series of vanes mounted on the circumference of a large wheel, as shown in Figure (10-4):

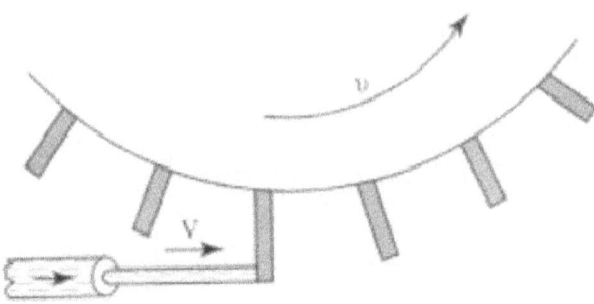

Fig (10-4) Jet of water impinging on a series of vanes

Let,

V = velocity of the jet in m/s,

a = cross-sectional area of the jet m^2, and

υ = velocity of the vanes as a result of the impact of the jet in m/s.

The jet of water, after impinging on the vanes, will be moving with a velocity of υ m/s. Since the force exerted by the jet is:

$$F = \text{mass of water flowing per sec x change of velocity}$$
$$= w \, a \, V/g \times (V - \upsilon) \qquad\qquad (i)$$

And,

$$\text{Work done by the jet} = \text{Force x distance}$$
$$= w \, a \, V/g \times (V - \upsilon) \times \upsilon \qquad\qquad (ii)$$

Therefore,

$$\text{Work done per kg of water} = L/g \, [(V - \upsilon) \times \upsilon] \qquad\qquad (iii)$$

$$\text{Energy of the jet water per kg of water} = V^2/2g$$

Therefore,

$$\text{Efficiency, } \eta = (\text{work done per kg of water}) / (\text{energy per kg of water})$$
$$= L/g \, [(V - \upsilon) \times \upsilon] / V^2/2g = [2 \, (V - \upsilon) \times \upsilon] / V^2$$

The efficiency is also equal to = (work done by the jet) / (energy of the jet)

10.1.5 Force of jet impinging on a fixed curved vane: Consider a jet of water, tangentially entering and leaving a fixed curve vane, as shown in Figure (10-5):

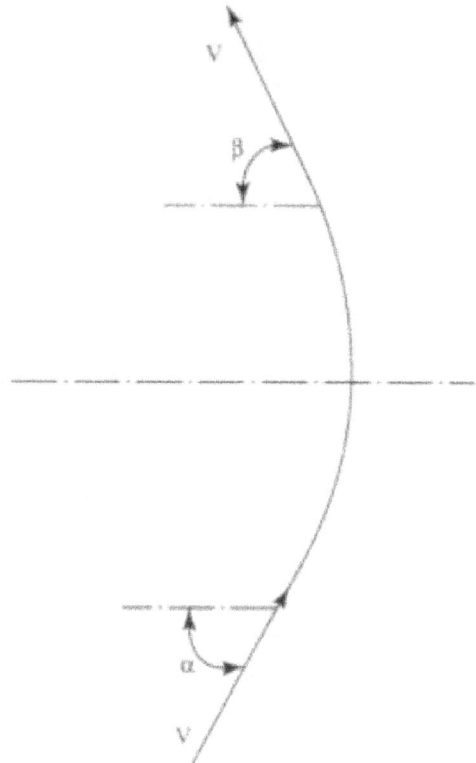

Fig (10-5) Water entering and leaving a vane

Let,

V = velocity of the jet,

a = cross-sectional area of the jet,

α = inlet angle of the jet, and

β = outlet angle of the jet.

The jet, while moving through the vane will exert some force on the vane. Tis force may be determined by finding out its components along the axial and perpendicular directions when normal of the vane.

The force of the jet along normal to the vane = mass of water flowing per second x change of velocity along normal to the vane.

$$= w\,a\,V/g\,(V\cos\alpha + V\sin\beta) \qquad (i)$$

And force of the jet along the perpendicular direction normal to the vane,

$$= w\,a\,V/g\,(V\cos\alpha - V\sin\beta)$$

10.1.6 Force of jet impinging on a moving vane: Consider a jet of water entering and leaving a moving curved vane as shown in Figure (10-6):

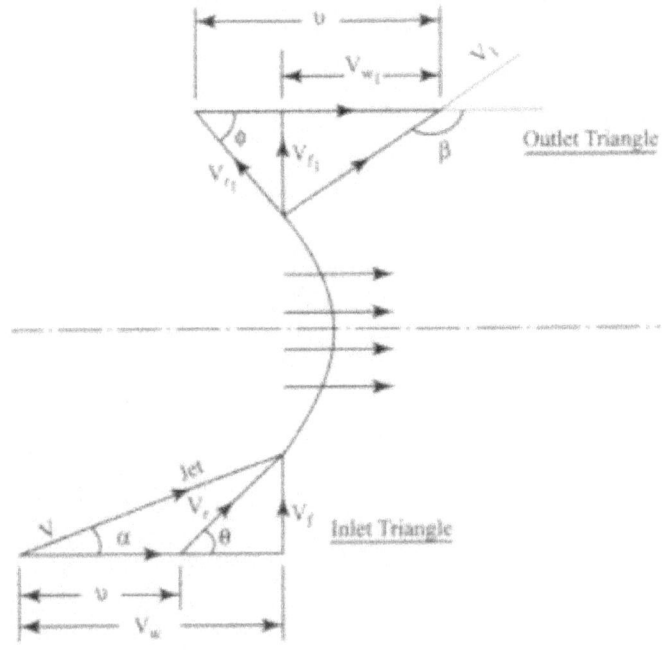

Fig (10-6) Moving vane

Let,

V   = jet velocity, at vane entry,

V_1 = jet velocity, at vane exit,

v, v_1 = vane velocity,

α   = angle with the direction of motion of the vane, at which the jet enters the vane,

β   = angle with the direction of motion of the vane, at which the jet leaves the vane,

V_r = relative velocity of the jet and the vane at entrance (it is the vectorial difference between V and v)

V_{r1} = relative velocity of the jet and the vane at exit (it is the vectorial difference between V_1 and v_1)

θ = angle which V_r makes with the direction of motion of the vane at inlet (known as vane angle at inlet)

Φ = angle which V_{r1} makes with the direction of motion of the vane at outlet (known as vane angle at outlet)

V_w= horizontal component of V i.e. a component parallel to the direction of motion of the vane (known as velocity of whirl at inlet) = V cosα

V_{W1}= horizontal component of V_1 i.e. a component parallel to the direction of motion of the vane (known as velocity of whirl at outlet) = V_1 cosβ

V_f = vertical component at V i.e. a component right angles to the direction of motion of the vane (known as velocity of flow at inlet) = V sinα, and

V_{f1} = vertical component of V_1 i.e. a component at right angles to the direction of motion of the vane (known as velocity of flow at outlet) = V_1 sinβ

Notations with suffix i.e. V, V_r, V_w, V_f stand for the inlet triangle and the notations with suffix i.e. V_1, V_{r1}, V_{W1}, V_{f1} stand for the outlet triangle. As the jet of water enters and leaves the vanes tangentially, therefore the shape of the vanes will be such that V_r and V_{r1} move along a tangent route at both inlet and outlet. It is thus obvious that the shape of the vanes is always designed according to the given data (i.e. first the triangle is drawn with the given data and then the vane is drawn in such a way, that V_r and V_{r1} are along the tangents to the vanes at inlet and outlet).

The relation between the inlet and outlet triangles is:

(a) $υ = υ_1$, and

(b) $V_r = V_{r1}$

Let,

a = cross-sectional area of the jet

The force of jet, in the direction of motion of the vane is:

$$F_x = \text{mass of water flowing per second x change of velocity of whirl}$$
$$= w\, a\, V/g\, (V_w - V_{W1}) \qquad\qquad (i)$$

And work done, in the direction of motion of the vane = Force x distance

$$= w\, a\, V/g\, (V_w - V_{W1}) \times υ \qquad (ii)$$

And work done per kg of water = L/g $(V_w - V_{W1})$ x υ $\qquad\qquad$ (iii)

10.2 Pressure of Water due to Deviated Flow

As a pipeline, carrying water, changes its direction from its straight path, the velocity of water flowing through the pipe will also change its value due to change in its direction. It is to be noted that diverted flow of water in pipes will cause some pressure on the pipe wall.

Consider pipeline carrying water and diverted from its straight path as shown in Figure (10-7):

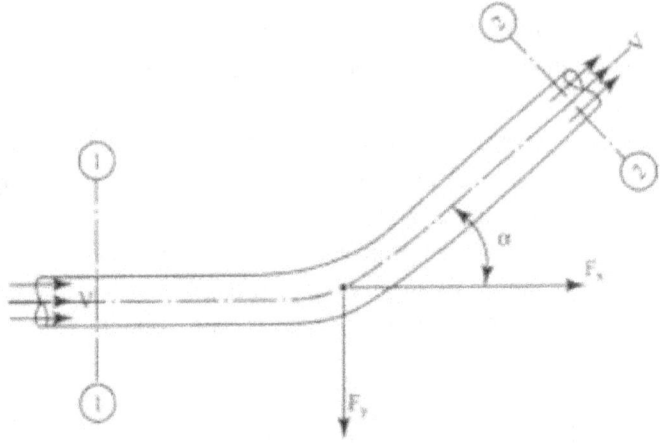

Fig (10-7) Pipeline deviated from its straight path

Let,

V = velocity of water in the pipe at section 1,

a = area of the pipe, and

α = angle through which the centre of pipe is deviated from its straight path.

Therefore,

$$\text{Velocity of water at section (2)} = V \cos \alpha$$

$$\text{Mass of water flowing at section (1)} = w\, a\, V/g$$

And momentum of the flowing water at section (1) in the X-X direction = mass x velocity

$$= (w\, a\, V/g) \times V$$
$$= (w\, a\, V^2) / g$$

Similarly, momentum of the flowing water at section (2) in the X-X direction,

$= (w\, a\, V/g) \times V \cos\alpha = (w\, a\, V^2 \cos\alpha) / g$ .... (mass of water at section (2) = $w\, a\, V/g$

Therefore,

Change in momentum $= [(w\, a\, V^2) / g] - [(w\, a\, V^2 \cos\alpha) / g] = [(w\, a\, V^2) / g]\,[1 - \cos\alpha]$

In accordance with Newton's second law of motion, the force is equal to the rate of change of momentum. Therefore, the force in X-X direction is:

$$F_x = [(w\, a\, V^2)\, \sin\alpha / g]\,[1 - \cos\alpha]$$

Similarly, it can be proved that the force in Y-Y direction is:

$$F_y = (w\ a\ V^2\ \sin\alpha)\ /\ g$$

The resultant force on the bend is:

$$R = \sqrt{(F_x^2 + F_y^2)}$$

If the resultant makes an angle Θ with X-X direction, then:

$$\tan\Theta = F_y\ /\ F_x$$

Note: sometimes the water flows under pressure 'p' through the pipe. In such a case force exerted because of pressure = p a

The component of this force in x-x direction = p a (1 - cosα)

And the component of this force in the Y-Y direction = p a sinα

Then, the total force in X-X direction = F_x + p a (1 - cosα)

And the total force in Y-Y direction = F_y + p a sinα

10.3 Jet Propulsion

As a jet strikes a plate, it exerts some force on the plate, and the force exerted is:

$$F = w\ a\ V^2\ /\ g$$

Where,
w = specific weight of the jet liquid,
a = area of the jet, and
V = velocity of the jet.

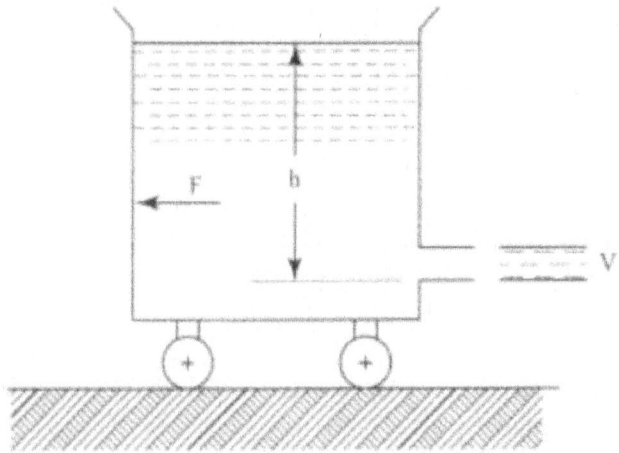

Fig (10-8) Jet discharging from a vessel

If such a jet is discharged from a vessel (which can move on its wheels); then according to Newton's third law of motion, the force of jet will tend to move the vessel in the opposite direction with the same force. Tis principle is known as: principle of jet propulsion and is used in driving ships and aeroplanes.

10.3.1 Propulsion of ships by water jets: The principle of jet propulsion is utilised in driving ships. The ship carries pumps, which take water from its surrounding. This water is discharge by forcing it through an orifice at the back of the ship. The efficiency of a ship depends on the direction of the inlet orifice. In general, a ship may have:

(i) its inlet orifices at right angles to the direction of its motion,
(ii) its inlet orifices facing the direction of motion.

10.3.2 Propulsion of ship having inlet orifices at right angles to the direction of its motion (i.e. orifices at mid-ship): Consider a ship having inlet orifices at right angles to the direction of its motion as shown in Figure (10-9):

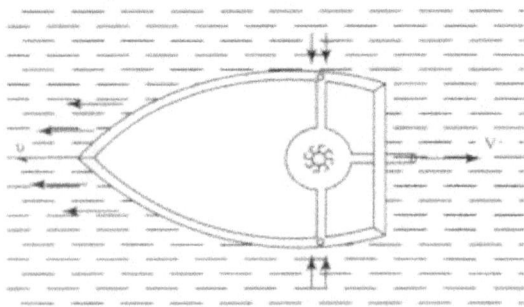

Fig (10-9) Ship with orifice at right angle to the direction of motion

Let,

V= velocity of the jet issuing from the ship,

a = area of the jet, and

v = velocity of the ship

Therefore, relative velocity of the jet and ship (V_r) will be equal to the vectorial difference of V and v. Since v is in the opposite direction of V, therefore their vectorial difference will be equal to (V + v):

$$V_r = V + v$$

Therefore,

Mass of the water flowing/sec = w a V_r / g

The velocity of the jet leaving the ship is equal to v.

Therefore,

Force exerted on the boat = mass of water flowing/sec x change of velocity

$$= w a V_r / g (V_r - v)$$

And,

work done by jet = Force x distance

$$= w a V_r / g (V_r - v) v$$

Therefore,

work done by the jet per kg of water = 1/g (V_r - v) v

Since, the energy supplied by the jet, per kg of water = V_r ^{2}/2g

Therefore,

Efficiency of propulsion η = work done by the jet per kg of water / energy of the jet per kg of water

$$= [1/g (V_r - v) v] / [V_r {}^2/2g]$$
$$= [2(V_r - v) v] / 2g \qquad (i)$$

The efficiency of propulsion will be maximum when 2(V_r - v) v or (V_r - v) v is maximum. Hence by differentiating this equation and equating the same to zero, we get:

$$dη/dv (V_r - v) v = 0$$
$$dη/dv (V_r v - v^2) = 0$$
$$V_r - 2 v = 0$$
$$v = V_r/2 \text{ and } V_r = 2 v$$

Substituting this value of V_r in equation (i) we get:

$$η = [2(2v - v) v] / 2((v)^2$$
$$= 2(v)^2/4(v)^2 = ½ = 0.5 = 50 \%$$

10.3.4 Propulsion of a ship having inlet orifice facing the direction of flow: Consider a ship having inlet orifice facing the direction of flow as shown in Figure (10-10):

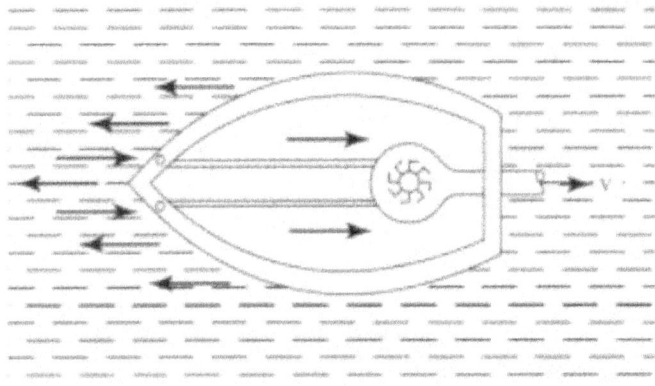

Fig (10-10) Ship having inlet orifice facing the direction of flow

Let,

V= velocity of the jet issuing from the ship,

a = area of the jet, and

v = velocity of the ship

Therefore,

Mass of water flowing/sec = w a V_r / g

The relative velocity of jet, after leaving the ship will be equal to v,

Therefore,

Force exerted on the boat = Mass of water flowing/sec x change of velocity

= w a V_r / g $(V_r - v)$

And work done by the jet = Force x distance

= w a V_r / g $(V_r - v)$ v

Therefore,

Work done by the jet per kg of water = $1/g$ $(V_r - v)$ v

The energy supplied by the jet per kg of water = V_r^2 /2g

When a ship moves in steady water, the water will enter into the orifice of the ship with a velocity equal to that of the ship (i.e. v). Therefore energy of the water entering the orifice per kg of water = $v^2/2g$

Net energy supplied by the jet = $(V_r^2 /2g - v^2/2g)$

Therefore,

Efficiency of propulsion, η = work done by the jet per kg of water / energy of the jet per kg of water

= $[1/g (V_r - v) v] / [(V_r^2 /2g - v^2/2g)]$

$$= [2 (V_r - \upsilon) \upsilon] / (V_r^2 - \upsilon^2]$$
$$= 2 \upsilon / (V_r^2 + \upsilon) \quad .....[V_r^2 - \upsilon^2 = (V_r - \upsilon)(V_r + \upsilon)]$$

Now power of the pump, $P = w\ a\ V_r/75\ (V_r^2\ /2g - \upsilon^2 /2g)$ HP

Check Your Knowledge

1) What do you understand by the term 'jet of water'? Write an expression for the force of jet on a fixed plate.

2) Show that the normal force exerted by a jet of water on an inclined plate is given by the relation,

$F = (w\ a\ V^2\ \sin\Theta)/g$

Where,

a = area of jet,

V = velocity of the jet, and

Θ = inclination of the plate with the jet.

3) Find an expression for the force of jet impinging on a moving plate and compare it with the force, when the same jet is impinging on a series of moving vanes.

4) Find an expression for the force, work done and efficiency of a moving curved vane.

5) Describe how ships move in sea using 'jet propulsion'.

11. Reciprocating Pumps

11.1 Introduction

A reciprocating pump, in its simplest form, consists of the following parts as shown in Figure (11-1):

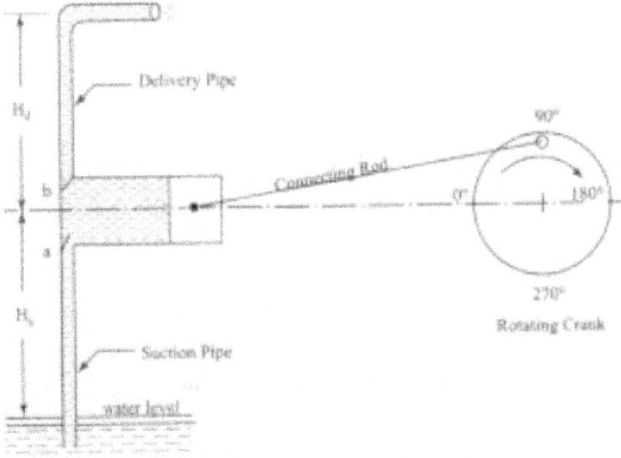

Fig (11-1) Reciprocating pump

1) A cylinder C, in which a piston P works. The movement of the piston is obtained by a connecting rod, which connects the piston and the rotating crank.

2) A suction pipe, connecting the source of water and the cylinder.

3) A delivery pipe, into which the water is discharged from the cylinder.

4) A valve a, which admits the flow from the suction pipe into the cylinder.

5) A valve b, which admits the flow from the cylinder into the delivery pipe.

During the suction stroke, the piston P moves towards the right creating vacuum in the cylinder. The vacuum causes the suction valve 'a' to open and the water enters the cylinder.

During the delivery stroke, the piston P moves towards the left increasing pressure in the cylinder. This increase in pressure causes the suction valve 'a' to close and delivery valve 'b' to open and the water is forced into the delivery pipe.

A reciprocating pump is also called a positive displacement pump, as it discharges a definite quantity of liquid during the displacement of its piston or plunger. That is why a reciprocating pump is ideally suitable for grouting operations in dam foundations.

11.2 Types of Reciprocating Pumps

The reciprocating pumps may be classified as stated below:

1. According to action of water:
 (a) single acting pump, and
 (b) double acting pump.

2. According to number of cylinders:
 (a) single cylinder pump,
 (b) double cylinder pump, and
 (c) triple cylinder pump, etc....

3) According to the existence of air vessels:
 (a) with air vessel, and
 (b) without air vessel.

11.3 Discharge of Reciprocating Pump

Consider a single acting reciprocating pump (i.e. a pump, in which the water is acting on one side of the piston only).
Let,
A = cross-sectional area of the piston,
L = length of the stroke or piston, and
N = number of revolutions per minute of the crank.

Then, the discharge of the pump, $Q = L \, A \, N/60$

If the pump is a double acting reciprocating pump (i.e. a pump in which the water is acting on both sides of the piston) the discharge is taken to be double the discharge than that of a single acting pump.

This is due to the fact that in a double acting pump the water is sucked on one side of the pistons and delivered from the other side during the same stroke. These two processes (i.e. suction on one side and delivery from the other) are reversed during the return stroke.

Therefore, the discharge of a double acting reciprocating pump is:

$$Q = 2 \text{ L A N}/60$$

11.3.1 Pump slip: The difference between the theoretical discharge and the actual discharge of a pump is known as 'slip of the pump'.

11.3.2 Negative slip of the pump: Sometimes the actual discharge of a reciprocating pump is more than the theoretical discharge. In such cases, the coefficient of discharge will be more than unity and the corresponding slip is known as 'negative slip' of the pump.

This happens when the suction pipe is long and delivery pipe is short and the pump is running at high speed. This causes the delivery value to open before completion of the suction stroke and some water is pushed into the delivery pipe before the piston commences its delivery stroke.

11.4 Power Required Driving a Reciprocating Pump

Consider a reciprocating pump, first withdrawing liquid (through the suction pipe) and then delivering the same (through the delivery pipe).
Let,
H_s = suction head of the pump in metres,
H_d = delivery head of the pump in metres, and
w = specific weight of the liquid.

Force on the piston in forward stroke = $w H_s A$ kg, and
Force on the piston in backward stroke = $w H_d A$ kg

Let,
Q = discharge of the liquid in m^3/s

Then,

Work done by the pump = $w Q (H_s + H_d)$ kg-m

Therefore,

Theoretical power required to drive the pump = $w Q (H_s + H_d)/75$ HP

The actual power, required to drive the pump will be more than the theoretical power due to various losses.

11.5 Effect of the Acceleration of Piston on the Indicator Diagram

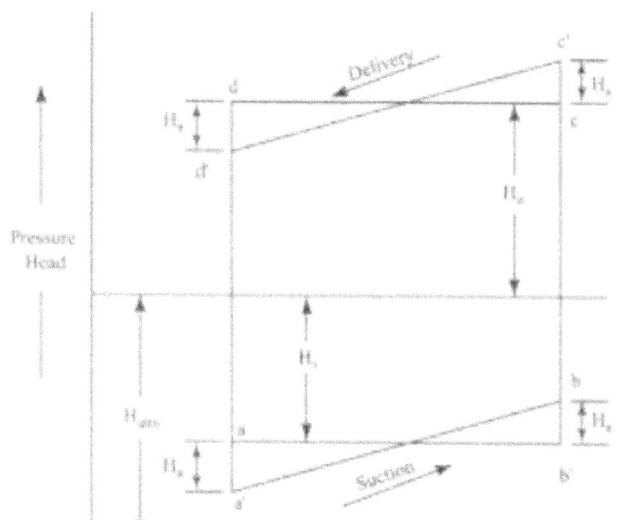

Fig (11-2) Indicator diagram

An acceleration pressure head is caused by the acceleration of the piston. At the beginning of the suction stroke, the pressure head is below the atmospheric pressure head by $(H_s + H_a)$, where H_a is the acceleration pressure head. In the middle of the suction stroke, the pressure head is below the atmospheric pressure head by H_s (as the acceleration pressure head $H_a = 0$, when $\theta = 90°$). At the end of the suction stroke, the pressure head is below the atmospheric pressure head by $(H_s - H_a)$.

Therefore, we can modify the indicator diagram for the suction stroke as shown in Figure (11-2) - a to $\grave{a}$ and b to $\grave{b}$. Similarly at the beginning of the delivery stroke, the pressure head is above the atmospheric pressure by $(H_d + H_a)$. In the middle of the delivery stroke, the pressure head is above the atmospheric pressure head by H_d. At the end of the delivery stroke, the pressure head is above the atmospheric pressure head by $(H_d - H_a)$. Therefore, the indicator diagram can be modified for the suction stroke as shown in Figure (11-2).

11.5.1 Effect of friction in the suction and delivery of pipes on the indicator diagram:
Whenever water is flowing through a pipe, there is always some loss of head due to friction, which offers resistance to the flow of water. Similarly, as the water is flowing through the suction and delivery pipes there will be a loss of head due to friction in both pipes.
Let,

A = area of the cylinder or bore,
d = diameter of the pipe,
a = area of the pipe,
ω = angular velocity of the rotating crank in radians/sec.
r = radius of the rotator crank,
L = length of pipe,
f = coefficient of friction, and
v = velocity of water in the pipe.

As the velocity of the piston at any instant $= \omega \, r \sin\omega t = \omega \, r \sin\Theta$

Therefore,

Velocity of water in the pipe at that instant, $v = A/a \; \omega \, r \sin\Theta$

The loss of head due to friction is:
$$H_f = (4fL v^2) / (2gd) = (4fL) / (2gd) \, (A/a \; \omega \, r \sin\Theta)^2$$

Following is discussion on the effect of pipe friction as reflected on the indicator diagram at the beginning, middle and end of the stroke:

1) At the beginning of the stroke, $\Theta = 0$. Therefore, the velocity of water in the pipe is zero, consequently, there is no loss of head due to friction.

2) At the middle of the stroke, $\Theta = 90°$. Therefore, $\sin\Theta = 1$, and the loss of head due to friction: $H_f = (4fL v^2) / (2gd) \, (A/a \; \omega \, r)^2$.

3) At the end of the stroke, $\Theta = 180°$. The velocity of water in the pipe is zero, consequently, there is no loss of head due to friction.

The effect of friction in the suction pipe, and the beginning of the suction stroke is that: the pressure head will be below the atmospheric pressure head by an amount of $(H_s + H_a)$, as H_f is zero. The pressure head in the middle of the suction stroke will be below the atmospheric pressure by the amount of $(H_s + H_f)$, because H_a is zero. At the end of the suction stroke, the pressure head is below the atmospheric pressure by $(H_s - H_a)$, because H_f is zero.

Therefore the modified indicator diagram for the suction is as shown in Figure (11-3):

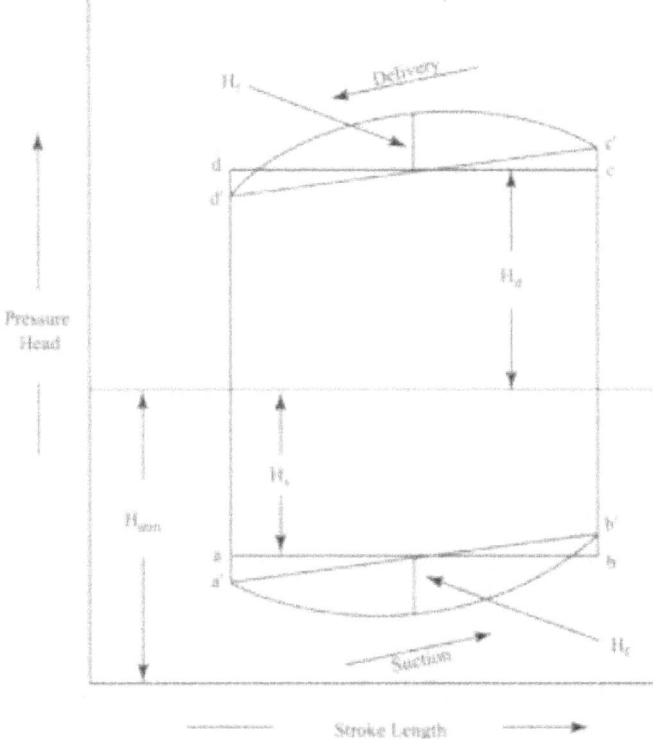

Fig (11-3) Modified indicator diagram

Similarly, at the beginning of the delivery stroke, the pressure head will be above the atmospheric pressure by $(H_d + H_a)$, as H_f is zero. In the middle of the delivery stroke, the pressure head will be above the atmospheric head by $(H_d + H_f)$ as H_a is zero. At the end of the delivery stroke, the pressure head will be above the atmospheric pressure by $(H_d - H_a)$, as H_f is zero. Therefore, the modified indicator diagram for the delivery stroke is as shown in Figure (11-3).

11.5.2 Effect of fitting air vessel to the suction and delivery pipes of a reciprocating pump: As air vessel is a cast iron closed chamber, having an opening at its base, through which water flows into the vessel, or from the vessel. The vessel is filled with compressed air.

The air vessels are fitted to both the suction and delivery pipes, close to the cylinder of the pump as shown in Figure (11-4):

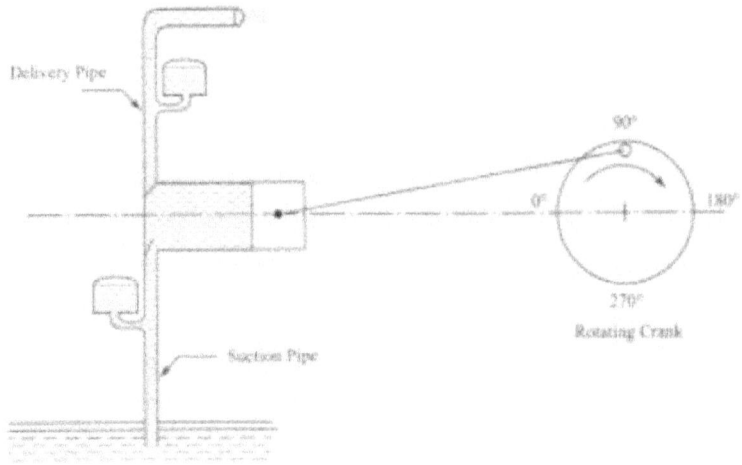

Fig (11-4) Reciprocating pump

The object behind fitting the air vessel is to obtain a uniform discharge from a reciprocating pump. Consider an air vessel fitted to the delivery pipe as shown in Figure (11-4) above. During the first half of the delivery stroke, the piston moves with acceleration thus forcing the water into the delivery pipe with a velocity more than the mean velocity. The excess flow of water, flows into the air vessel thus compressing the air inside the vessel. During the second half of the delivery stroke, the piston moves with retardation thus forcing the water into the delivery pipe, with a velocity less than the mean velocity. The water, stored into the air vessel, starts flowing into the delivery pipe adjusting the deficiency in flow.

Thus, the discharge in the delivery pipe, behind the air vessel, is more or less uniform. For practical purposes, the velocity in the delivery pipe, behind the air vessels, is taken as uniform. Similarly, on the suction side the water flows from the suction pipe into the air vessel (during the first half of the suction stroke) and then from the air vessel to the cylinder (during the second half of the suction stroke). Thus, for practical reasons, the velocity in the suction pipe of a reciprocating pump, up to the air vessel, is taken to be as uniform.

11.5.3 Maximum speed of the rotating crank with air vessel:

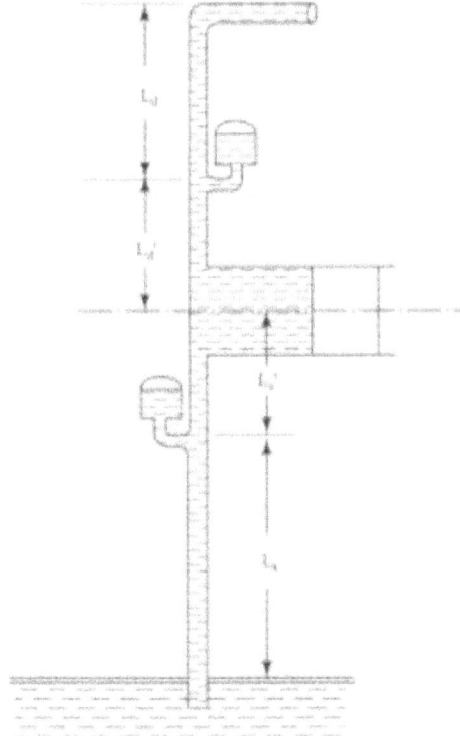

Fig (11-5) Reciprocating pump with air vessels

By fitting air vessels, the velocity of water in the suction pipe up to the air vessel (i.e. for a length of L_s)   is uniform. The acceleration and retardation to the velocity of water will take place in the suction pipe beyond the air vessel (i.e. for a length of L_s').

Similarly, acceleration and retardation to the velocity of water will take place in the delivery pipe up to the air vessel (i.e. for a length of L_d). The velocity of water in the delivery pipe beyond the air vessel (i.e. length of L_d) is constant.

Thus, to find out the maximum speed of the crank, there should be a limit to separation head:

$$H_{sep} = H - (H_s + H_a \text{ for } L_s' + H_f \text{ for } L_s)$$

The constant velocity of water, in the suction or delivery pipe, is found by dividing the discharge of the pump by the area of the respective pipe.
Let,
L = length of the stroke,
A = area of the piston,

N = speed of the pump in r.p.m,
ω = angular velocity of the crank,
a = area of the pipe,
r = radius of the crank, and
v = velocity of water in the pipe.

a) For single acting pump:
The discharge of a single acting pump is:

$$Q = LAN / 60$$

And,
Velocity of water is:

$$v = Q/a = LAN/60 \text{ x } a = A/a \text{ x } LN/60$$

Substituting, L = 2r, $\omega = 2\pi N/60$ or, N = $60\omega/2\pi$ in the above equation we get:

$$v = A/a \text{ x } 2r \text{ x } [(60\omega/2\pi)/60] = A/a \text{ x } \omega r/\pi$$

b) For double acting pump:
The discharge of a double acting pump is:

$$Q = 2LAN / 60$$

And,
The velocity of water is:

$$v = Q/a = 2LAN / 60 \text{ x } a = 2A/A \text{ x } LN/60$$

Substituting, L = 2r, $\omega = 2\pi N/60$ or, N = $60\omega/2\pi$ in the above equation we get:

$$v = 2A/a \text{ x } 2r \text{ x } [(60\omega/2\pi)/60] = A/a \text{ x } \omega r/\pi$$

11.5.4 Work done against friction, without air vessels: Since there is no loss of head due to friction at the beginning and end strokes, hence the maximum loss of head is in the middle stroke. The indicator diagram shows that the pressure head due to friction is a parabola.
Let,
A = area of the cylinder,
a = area of the pipe,
ω = angular velocity of the crank,
r = radius of the crank,
L = length of the pipe,
d = diameter pf the pipe, and
f = coefficient of friction.

At the middle of the stroke, the speed of water is:

$$\upsilon = A/a \ \omega \ r$$

And,

Loss of head due to friction is:

From the cylinder to the air vessel is subjected to acceleration water $H_f = 4fL\upsilon^2/2gd =$
$$[4fL(A/a \ \omega \ r)^2] \ 2gd$$

Since the area of a parabola = 2/3 x base x height
Therefore,
$$\text{Work done per stroke} = 2/3 \ x \ [4fL/2gd \ (A/a \ \omega \ r)^2]$$

11.5.5 Work saved, against friction, by fitting air vessel: The work saved against friction by fitting an air vessel, is found by finding the work done against friction without air vessels and then subtracting from it the work done against friction with air vessels.

11.5.6 Flow of water, into and from the air vessel, fitted to the delivery pipe of a single acting reciprocating pump: As the velocity of water in the delivery pipe, beyond the air vessel, is constant, and the velocity of water from the cylinder to the air vessel is subjected to acceleration and retardation, hence our discussion will take into consideration the rate of flow of water into and from the air vessel.

Consider a single acting reciprocating pump, fitted with air vessels on both the suction and delivery pipes, and examining the discharge from the cylinder to the delivery pipe, we get:

The velocity of water in the delivery pipe from the cylinder up to the air vessel= $A/a \ \omega \ r \ sin\Theta$
Therefore,
$$\text{Discharge from the cylinder} = a \ x \ A/a \ \omega \ r \ sin\Theta = A \ \omega \ r \ sin\Theta \qquad (i)$$

The velocity of water in the delivery pipe, beyond the air vessel = $A/a \ x \ \omega r/\pi$
Therefore,
$$\text{Discharge in the delivery pipe beyond the air vessel} = a \ x \ (A/a \ x \ \omega r/\pi) = A \ \omega r/\pi \quad (ii)$$

The difference between the above two discharges will be that discharge into or from the air vessel:
Therefore,
Discharge into the air vessel, Q = discharge from the cylinder - discharge beyond air vessel
$$= A \ \omega r \ sin\Theta - A \ \omega r/\pi = A\omega r \ (sin\Theta - L/\pi) \qquad (iii)$$

If the above equation works out to be positive, it means that the discharge is taking place into the air vessel. But if this equation works out to be negative, it means that the discharge is taking place from the air vessel.

Note: If we consider the flow, into or from the air vessel fitted to the suction pipe, then the above condition is reversed i.e. if equation (iii) above works out to be positive, the discharge is taking place from the air vessel. But if the equation works out to be negative, the discharge is taking place into the vessel.

11.5.7 Flow of water, into and from the air vessel, fitted to the delivery pipe of a double acting reciprocating pump: Consider a double acting reciprocating pump, fitted with air vessels on both the suction and the delivery pipes. Examining the discharge from the cylinder to the delivery pipe:

The water velocity in the delivery pipe from the cylinder up to the air vessel = $A/a \, \omega \, r \sin\Theta$
 Therefore,

$$\text{Discharge from the cylinder} = a \times A/a \, \omega \, r \sin\Theta \ = A \, \omega \, r \sin\Theta \tag{i}$$

The velocity of water in the delivery pipe, beyond the air vessel = $(2A/a) \, (\omega \, r/\pi)$
Therefore,
 Discharge in the delivery pipe beyond the air vessel = $a \times (2A/a) \, (\omega \, r/\pi) = 2A(\omega \, r/\pi)$ (ii)

The difference between the above two discharges is the discharge into or from the air vessel.

$$Q = \text{Discharge from the cylinder - Discharge beyond the air vessel}$$
$$= A \, \omega \, r \sin\Theta - 2A \, (\omega \, r/\pi)$$
$$= A \, \omega \, r \, (\sin\Theta - 2/\pi) \tag{iii}$$

If the above equation works out to be positive, it means that the discharge is taking place into the air vessel. But if this equation works out to be negative, it means that the discharge is taking place from the air vessel.

Note: If we consider the flow into or from the air vessel fitted to the suction pipe, then the above condition is reversed i.e. if the equation (iii) above works out to be positive, the discharge is taking place from the air vessel. But if this equation works out to be negative, the discharge is taking place into the air vessel.

Check Your Knowledge

1) Why a reciprocating pump is called a positive displacement pump?

2) Distinguish between coefficient of discharge and slip of a reciprocating pump.

3) What is the function of non-return valves in a reciprocating pump?

4) Explain the working principle of a reciprocating pump with sketches.

5) What is an indicator diagram of a reciprocating pump? What useful information does it give?

6) What are the factors, which influence the speed of a reciprocating pump?

7) Explain the function of air vessels in a reciprocating pump.

8) Drive an equation for the work saved by fitting air vessels in a reciprocating pump.

9) The work saved by fitting an air vessel on a single acting reciprocating pump is more than in a double acting reciprocating pump. Why?

12. Turbines

12.1 Classification of Turbines

Turbine may be broadly classified into the following two main groups:

a) Impulse or velocity turbines, and
b) Pressure or reaction turbines

12.1.1 Impulse turbines: In such turbines, the entire available energy of the water is first converted into kinetic energy, by passing it through the nozzle, which are kept close to the runner. The water enters the running wheel in the form of a jet, which impinges on the buckets, fixed to the outer periphery of the wheel.

The jet of water impinges on the buckets with a high velocity and after flowing over the vanes, leaves with low velocity; thus imparting energy to the runner. The pressure of water, both at entering and leaving the vanes, is atmospheric. The commonest example of an impulsive turbine is Pelton wheel.

12.1.2 Pelton wheel: A Pelton wheel is an impulsive turbine, used for high heads of water. It has a wheel to which a number of buckets or cups are mounted round the periphery as shown in Figure (12-1):

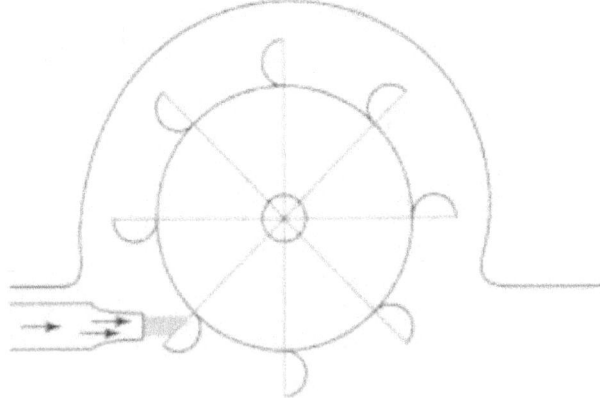

Fig (12-1) Pelton wheel

The cups mounted on the periphery, resembles a hemispherical cup or bowl with a dividing wall in the centre of the cup. This dividing wall is called splitter and the runner vane or cup is known as a bucket.

The jet of water, issuing from the nozzle, strikes the bucket at its splitter. The splitter then splits up the jet in two parts. One part, of the jet, runs over the inside surface of one portion of the vane and leaves it at its extreme edge. The other part, of the jet runs over the inside surface of the other portion of the vane and leaves it at its extreme edge as shown in Figure (12-2):

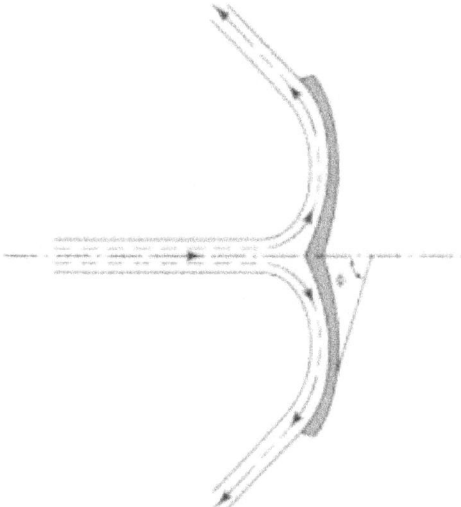

Fig (12-2) Bucket splitter

The mid-point of the bucket, where the jet strikes the splitter and gets divided, forms the inlet tip and the two extreme edges, where the divided jet leaves the bucket, from the two outlet tips.

The inlet velocity triangle is drawn at the splitter (which will be a straight line) and the outlet velocity triangle is drawn at any one of the outer tips of the hemispherical bucket as shown in Figure (12-3):

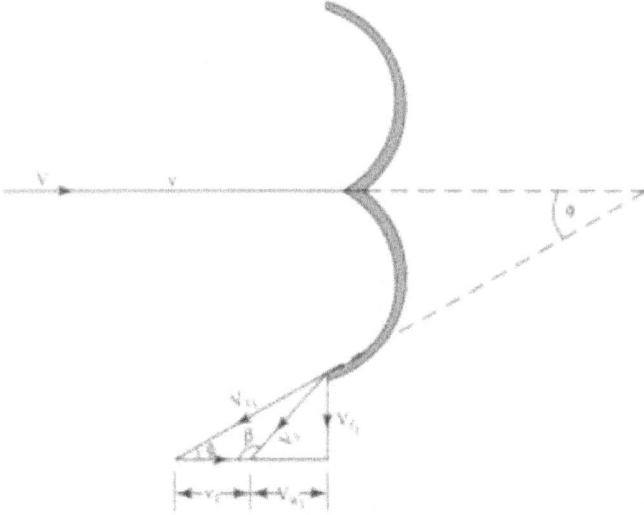

Fig (12-3) Mid-point of buckets with formation of inlet tip

Let,

V = absolute velocity of the entering water,

V_1 = absolute velocity of the leaving water,

D = diameter of the wheel,

d = diameter of the nozzle,

N = revolutions of the wheel in r.p.m,

υ = tangential velocity of buckets, (also known as peripheral velocity of the wheel),

V_r = relative velocity of water to bucket at inlet,

V_{r1}= relative velocity of water to the bucket at outlet,

V_f = velocity of flow at inlet,

V_{f1}= velocity of flow at outlet,

φ = angle of the blade tip at outlet, and

H = total head of water, under which the wheel is working.

Since the inlet triangle is a straight line, therefore velocity of whirl at inlet is:

$$V_W = V, \quad \text{and} \quad V_r = V - \upsilon$$

From the outlet triangle, we find that velocity of whirl at outlet is:

$$V_{W1} = V_{r1} \cos\varphi - \upsilon = (V - \upsilon) \cos\varphi - \upsilon$$

Since the force per kg of water = change of velocity of whirls = $V_W - V_{W1}$ (+ sign of V_W is taken as the V_W is negative in Pelton wheels)

And,

Work done per kg of water = Force x distance = $1/g \ V_W \ \upsilon + V_{W1} \ \upsilon_1 = V_W \ \upsilon/g + V_{W1} \ \upsilon/g$ (since $\upsilon_1 = \upsilon$)

$$= V_W \ \upsilon/g + (V_{r1} \cos\varphi - \upsilon) \ \upsilon/g \qquad (V_{W1} = V_{r1}$$

$\cos\varphi - \upsilon)$

$$= \upsilon/g\{V_W + [(V - \upsilon) \cos\varphi - \upsilon]\} \qquad (\text{since } V_{r1} =$$

$V_r = V - \upsilon)$

$$= \upsilon/g \ (V + V \cos\varphi - \upsilon \cos\varphi - \upsilon) \qquad (\text{since } V_W =$$

$V)$

$$= \upsilon/g \ [V(1 + \cos\varphi) - \upsilon \ (1 + \cos\varphi)]$$
$$= \upsilon \ (V - \upsilon) \ (1 + \cos\varphi) \ /g$$

As the hydraulic efficiency, η_h = work done per kg of water / Energy supplied per kg of water

$$= [\upsilon \ (V - \upsilon) \ (1 + \cos\varphi) \ /g] \ / \ V^2/2g$$
$$= 2 \ \upsilon \ (V - \upsilon) \ (1 + \cos\varphi) \ /V^2$$

Differentiating the numerator, of the above equation, with respect to υ and equating it to zero for maximum efficiency (as the maximum efficiency occur when the numerator will be maximum) we get:

$$dE/d\upsilon \ 2 \ \upsilon(V - \upsilon)(1 + \cos\varphi) = 0$$

$$dE/d\upsilon \; [(2V\upsilon - 2\upsilon^2)(1 + \cos\varphi) = 0$$
$$2V - 4\upsilon = 0$$

Or,

$$\upsilon = V/2$$

This means that the velocity of the wheel, for maximum hydraulic efficiency, should be half of the jet velocity.

Therefore,

Maximum work done/kg of water $= \upsilon \, (V - \upsilon) \, (1 + \cos\varphi) \, /g$

$\qquad = [(\upsilon/2)(V - \upsilon/2)(1 + \cos\varphi)] \, /g$ $\qquad\qquad$ (substituting $\upsilon = V/2$)

$\qquad = V^2/4g \, (1 + \cos\varphi)$

And,

Maximum hydraulic efficiency, max $\eta_h = [V^2/4g \, (1 + \cos\varphi)] \, / \, V^2/2g$

$\qquad\qquad\qquad = (1 + \cos\varphi) \, / \, 2$

Note:

1) The efficiency is maximum when $\cos\varphi = 1$ i.e. $\varphi = 180°$. But in actual practice the jet is deflected through an angle $160°$ to $165°$ only since the jet if made to deflect through $180°$, the water discharged from one bucket will then have an impact on the bucket in front of it.

2) In actual practise, the maximum efficiency takes place when the velocity of wheel is 0.46 times the velocity of the jet.

12.1.3 Efficiencies of an impulse turbine: In general, the term efficiency is defined as the work done to the energy supplied. An impulse turbine has also the following efficiencies:

1) Hydraulic efficiency,
2) Mechanical efficiency, and
3) Overall efficiency

12.1.4 Hydraulic efficiency: Is the ratio of work done, on the wheel, to the energy of the jet. The hydraulic efficiency of a turbine is:

$\eta_h = $ [Work done/kg of water]/[Energy of the jet/kg]

$\qquad = [\upsilon(V - \upsilon)(1 + \cos\varphi) \, /g] \, / \, V^2/2g$

$\qquad = 2 \, \upsilon(V - \upsilon) \, (1 + \cos\varphi) \, / \, V^2$

12.1.5 Mechanical efficiency: Is the ratio of actual work available at the turbine to the energy impacted to the wheel.

12.1.6 Overall efficiency: Is a measure of the performance of a turbine and is the ratio of power produced by the turbine to the energy actually supplied to the turbine, i.e.

$$\eta_0 = \eta_h \times \eta_m = P / (WQh/75)$$

12.1.7 Power produced by an impulse turbine: As some work is done per kg of water, when the jet strikes the buckets of an impulse turbine, hence if we know the quantity of water in kg, flowing through the jets per second, we can find out the amount of work done per second. The horse power produced by the turbine may then be easily found using the relation:

P = [Work done/kg of water x weight of water in kg flowing/sec] / 75

12.1.8 Governing of an impulse turbine (Pelton Wheel): In practice, load on the generator (which is coupled to an impulse turbine) fluctuates from time to time.

This fluctuating load, on the generator, has some effect on the turbine as the generator is directly coupled to the turbine. The change of load on the turbine, will change its speed and rate of flow.

In order to have a high efficiency at different loads, the speed of the turbine must be kept constant, as far as possible. The process of providing constant speed and regulating the rate of flow (according to the change of load), is known as governing of the turbine. Though there are many methods of governing an impulse turbine, the servomotor method or Relay cylinder method is commonly used, which is discussed below:

The servomotor method is a mechanism consisting of the following parts as shown in Figure (12-4):

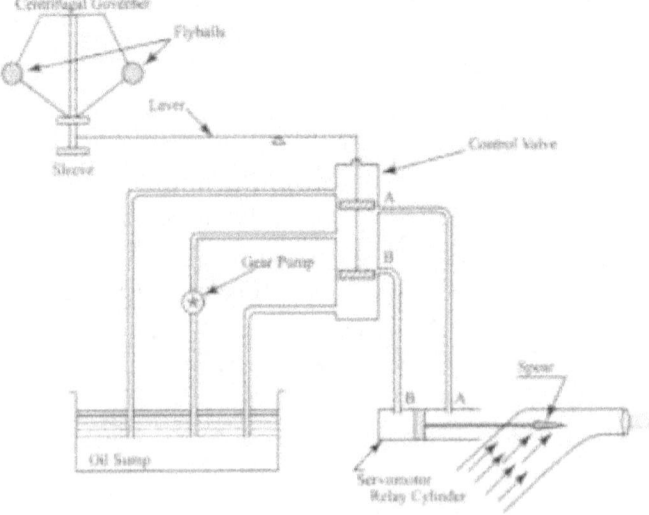

Fig (12-4) Servomotor

1) Centrifugal governor,
2) Control valve,
3) Servomotor,
4) Gear pump,
5) Oil pump,
6) Spear or needle, and
7) A set of pipes connecting oil sump with control valve and control valve with relay cylinder.

The centrifugal governor is driven from the main shaft of the turbine either by belt or gear arrangement. The control valve controls the direction of flow of the liquid (which is pumped by gear pump from the oil sump) either in pipe AA or BB. The servomotor or relay-valve has a piston (whose motion, towards left or right, depends upon the pressure of the liquid flowing through the pipes AA or BB) which is connected to a spear or needle reciprocating inside a nozzle as shown in Figure (12-4).

When the turbine is running at its normal speed, the positions of piston (in a servomotor or relay cylinder), control valve and fly balls of centrifugal governor will be in their normal positions as shown in Figure (12-4). The oil pumped by the gear pump into the control valve will return to the oil sump when pipes AA and BB are closed by the two wings of the control valve.

As the load on the turbine increases, the turbine speed will decrease. This decrease in speed of the turbine runner will also decrease the speed of the centrifugal governor; as a result of which, the fly balls will descend decreasing their amplitude (due to decrease in centrifugal force). This descends of the fly balls, will also cause the sleeves to descend, as they are connected to the central vertical bar of the centrifugal governor. This downward movement of the sleeve will raise the control valve rod (as the sleeve is connected to the control valve rod through a lever pivoted on a fulcrum).

A slight upward movement of the control valve rod will open the way to pipe AA (still keeping pipe BB closed). Now the oil (under pressure) will rush from the control valve to the right side of the piston in the servomotor through pipe AA. This oil, under pressure, will move the piston and spear towards the left, which will open more area of the nozzle, controlling the flow to the turbine.

The increase in the area of flow will increase the rate of flow, as a result of which there will be an increase in the speed of the turbine. When the speed of the runner will come up to the normal speed, fly balls and sleeve will move upward as the control valve rod will occupy its normal position.

It may be noted that when the load on the turbine decreases, its speed will increase. As a result of this, the fly balls will move upward (due to increase in centrifugal force) and sleeve will also move upward. This will push the control valve downwards, and the downward movement of the control valve rod will open pipe BB (still keeping pipe AA closed).

The oil (under pressure) will rush from the control valve to the left side of the piston in the servomotor through the pipe BB. This oil, under pressure, will move the piston and spear towards the right, which will decrease the area of the nozzle and ultimately decrease the rate of flow. This decrease in the rate of flow will decrease the speed of the turbine until the speed, once again, comes down to its normal speed.

12.2 Reaction Turbines

In a reaction turbine, the water enters the wheel under pressure and flows over the vanes. As the water is under pressure, the wheel of the turbine runs full and may be submerged below the tail race or may discharge into the atmosphere. The pressure head of water, while flowing over the vanes, is converted into velocity head and is finally reduced to the atmospheric pressure, before leaving the wheel.

A pressure or reaction turbine consists of fixed guide blades which guide the water to the moving vanes at the correct angle' so that the entry of water is shock less (this is done by adjusting the relative velocity of jet and vanes, tangentially to the vanes). The moving blades rotate due to energy available from the water passing through them. The wheel is surrounded by watertight casing.

12.2.1 Classification of reaction turbines: The reaction turbines may be classified into the following three types, depending on the direction of flow of water through the wheel:

(i) Radial flow turbines,
(ii) Axial flow turbines, and
(iii) Mixed flow turbines.

12.2.2 Radial flow turbines: In such turbines the flow of water is radial (i.e. along the radius of the wheel). The radial flow turbines may be further subdivided into the following two classes:

(a) Inward flow turbine:
In such turbines the water enters the wheel at the outer periphery and then flows inwards (i.e. towards the centre of the wheel).

(b) Outward flow turbine:
In such turbines, the water enters at the centre of the wheel and then flows outwards (i.e. towards the outer periphery of the wheel).

12.2.3 Axial flow turbines: In such turbines, the water flows parallel to the axis of the wheel. Such turbines are sometimes called parallel flow turbines.

12.2.4 Mixed flow turbines: In such turbines the flow is partly radial and partly axial.

12.2.5 Inward flow reaction turbine: The inward flow reaction turbine is that reaction turbine in which the water enters the wheel at the outer periphery and then flows inward over the vanes (i.e. towards the centre of the wheel) as shown in Figure (12-5):

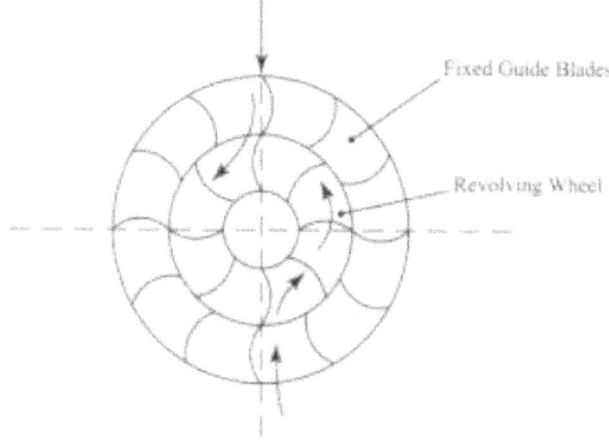

Fig (12-5) Inward flow reaction turbine

An inward flow reaction turbine, in its simplest form, consists of fixed guide blades, which guide the water into the revolving wheel at the correct angle, i.e. for the shockless entry of water. (This is done by adjusting the vane angle tangentially to the relative velocity of the water and the revolving wheel). The water, while passing over the vanes, exerts some force on the revolving wheel, to which the vanes are fixed. This force causes the wheel to revolve.

It may be noted that whenever the load on the turbine is decreased, it causes the shaft to rotate at a higher speed. The centrifugal force, which increases due to the higher speed, tends to reduce the quantity of water flowing over the vanes and thus the velocity of water at the entry is also reduced. It will ultimately tend to reduce the power produced by the turbine. This is the advantage of an inward flow reaction turbine where it can adjust automatically according to the required load on the turbine. The highest efficiency is obtained, when the velocity is as small as possible.

The efficiency of the power developed by the turbine may be found out by drawing the inlet and outlet velocity triangles, as shown in Figure (12-6):

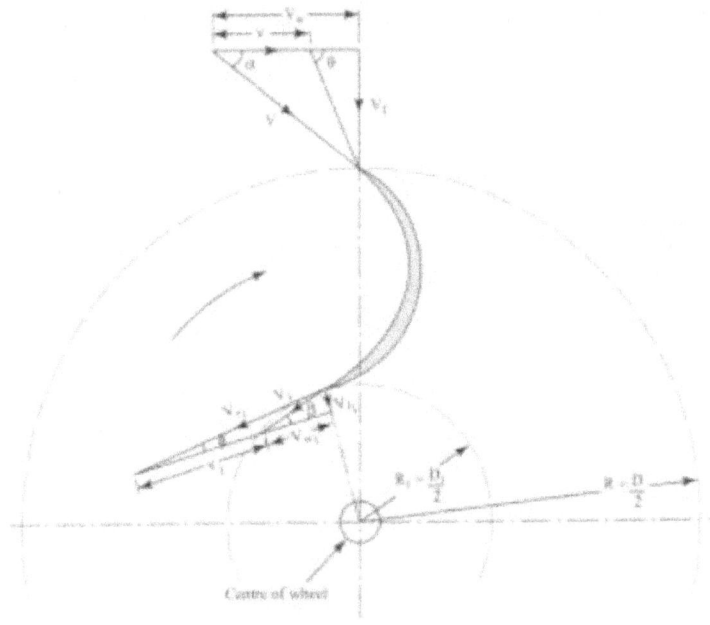

Fig (12-6) Inlet and outlet velocity triangles

Let,

V = absolute velocity of the entering water,

V_1 = absolute velocity of the leaving water,

D = outer diameter of the wheel,

D_1= inner diameter of the wheel,

N = revolution of the wheel per minute,

υ = tangential velocity of wheel at inlet (also known as peripheral velocity at inlet) = $\pi DN/60$,

υ_1 = tangential velocity of wheel at outlet (also known as peripheral velocity at outlet) = $\pi D_1 N/60$

V_r = relative velocity of water to the wheel at inlet,

V_{r1}= relative velocity of water to the wheel at outlet,

V_f = velocity of flow at inlet,

V_{f1}= velocity of flow at outlet,

α = angle, at which the water enters the wheel (also known as guide blade angle),

β = angle at which the water leaves the wheel,

Θ = angle of the blade tip at inlet (also known as vane angle at outlet),

H = total head of water under which the turbine is working, and

W = weight of the water entering the wheel in kg/sec.

From inlet triangle, we have:

$$V_w = V \cos\alpha$$

And,

$$V_f = V \sin\alpha$$

From outlet triangle, we have:

$$V_{w1} = V_1 \cos\beta$$

And,

$$V_{f1} = V_1 \sin\beta$$

As the force per kg of water = change of velocity of whirl = V_w - V_{w1}
And,
Work done per kg of water = Force x distance = (velocity of whirl at inlet x tangential velocity of wheel at inlet) = (velocity of whirl at outlet x tangential velocity of wheel

$$= 1/g\ (V_w\ \upsilon - V_{w1}\ \upsilon_1) = (V_w\ \upsilon/g) - (V_{w1}\ \upsilon_1/g) \qquad \text{(i)}$$

Energy lost per kg of water passing through the wheel = H - $V_1^2/2g$ (ii)

If there is no other loss of energy, then

$$(V_w\ \upsilon/g) - (V_{w1}\ \upsilon_1/g) = H - V_1^2/2g$$

If the discharge of the turbine is radial: i.e. $\beta = 0$, $V_{w1} = 0$, and $V_1 = V_{f1}$

Hence,

$$\text{Work done per kg of water} = V_w\ \upsilon/g \qquad \text{(since, } V_{w1} = 0)$$

And,

$$V_w\ \upsilon/g = H - V_1^2/2g = H - V_{f1}^2/2g \qquad \text{(since, } V_1 ⁻ V_{f1})$$

12.2.6 Outward flow reaction turbine:

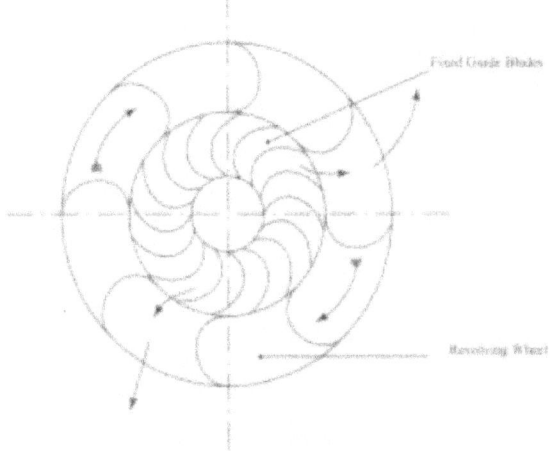

Fig (12-7) Reaction turbine - outward flow

An outward flow reaction turbine is that reaction turbine in which the water enters at the centre of the wheel and then flows outwards over the vanes (i.e. towards the outer periphery of the wheel).

An outward flow reaction turbine, in its simplest form, consists of fixed guide blades which guide the water to enter into the revolving wheel at the correct angle i.e. for shockless entry of water (this is done by adjusting the vane angle tangentially to the relative velocity of water and the revolving wheel).

The water, while flowing over the vanes, exerts some force on the revolving wheel to which the vanes are fixed. This force causes the wheel to revolve. The only difference between the inward flow reaction turbine and an outward flow reaction turbine is that in case of an inward flow reaction turbine, the revolving wheel is inside the fixed guide blades as shown earlier; whereas in case of an outward flow reaction turbine the revolving wheel is outside the fixed guide blades as shown in Figure (12-7).

Whenever the load on the turbine is decreased it causes the shaft to rotate at a high speed. The centrifugal force, which increases due to the higher speed, tends to increase the quantity of water flowing over the vanes and thus the wheel tends to run faster. This is the only disadvantage of an outward flow reaction turbine. Thus every outward flow reaction turbine has to be governed by a turbine governor.

All the notations for an outward reaction turbine are the same as that for inward flow reaction turbines. The inner diameter of the wheel will be denoted by D (i.e. diameter at inlet) and the outer diameter will be denoted by D_1 (i.e. diameter at outlet). All the relations, for finding out the various angles and other data will also hold good for an outward turbine.

The efficiency or power developed by the turbine may be found by drawing the inlet and outlet velocity triangles as usual as shown in Figure (12-8):

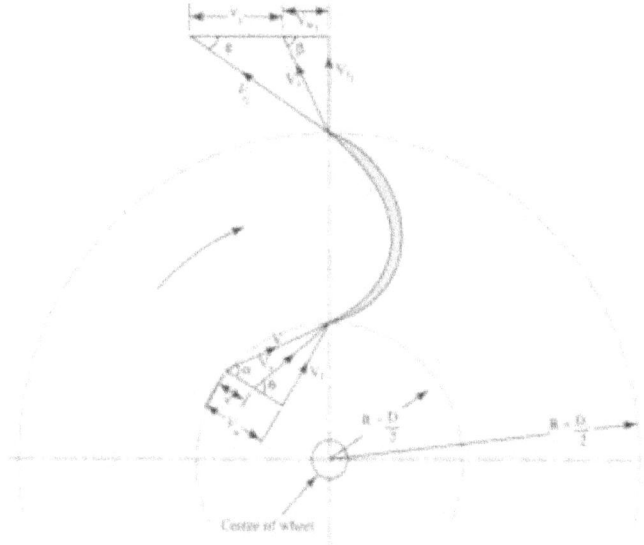

Fig(12-8) Inlet and outlet velocity triangles for outward reaction turbines

12.2.7 Discharge of a reaction turbine: The discharge of a reaction turbine is found out either from the gross energy supplied to the turbine or from the actual velocity of flow at inlet or outlet.

From the gross energy supplied to the turbine:
Let,
H = head of water supplied in meters, and
Q = discharge through the turbine in kg/sec or litres/sec.

Hence, the gross power supplied to the turbine = wQH/75

From the velocity of flow:
Let,
V_f = velocity of flow at inlet,
D = diameter of the wheel at inlet,
b = breadth of the wheel at inlet.

Water entering the wheel,
$$Q = \pi \, D \, b \, V_f \qquad\qquad (i)$$

Similarly, water leaving the wheel,
$$Q = \pi \, D_1 \, b_1 \, V_{f1} \qquad\qquad (ii)$$

As the water entering the wheel is equal to the water leaving the wheel, therefore,

$$\pi \, D \, b \, V_f = \pi \, D_1 \, b_1 \, V_{f1}$$

12.2.8 Efficiencies of a reaction turbine: In general, the term efficiency may be defined as the ration of work done to the energy supplied. Following are three types of efficiencies for a turbine:

(1) Hydraulic efficiency,
(2) Mechanical efficiency, and
(3) Overall efficiency.

12.2.9 Hydraulic efficiency: It is the ratio of work done on the wheel to the head of water (or energy) actually supplied to the turbine:

$$\eta_h = (\text{Work done / kg of water}) / H$$
$$= (V_w \, \upsilon/g - V_{w1} \, \upsilon_1/g) / H$$

If the discharge through the wheel is radial, then the velocity of whirl at outlet, i.e. $V_{w1} = 0$
Therefore,

$$\eta_h = V_w \, \upsilon/g \, H$$

12.2.10 Mechanical efficiency: Experience shows that all the energy supplied to the wheel does not result as useful work. But part of it is lost in overcoming friction on the bearings, and another part is lost in the governor. Thus the mechanical efficiency of a turbine: is the ratio of actual work available at the turbine to the energy supplied to the wheel.

The energy supplied to the wheel (in case of radial discharge) $= (V_w \, \upsilon/g) \times w \, Q$
Where,
Q = discharge of the turbine in m^2/sec
Therefore,

$$\text{Power supplied to the wheel} = (V_w \, \upsilon/g) \times w \, Q/75$$

Let,
P = Power available at the turbine (also known as brake horse power)
Therefore,

$$\eta_m = P / [(V_w \, \upsilon/g) \times w \, Q/75]$$

12.2.11 Overall efficiency of a turbine: is a measure of performance of a turbine. It is the ratio of power produced by the turbine to the energy actually supplied to the turbine:

$$\eta_0 = \eta_h \times \eta_m = (V_w \, \upsilon/g) \times P / [(V_w \, \upsilon/g) \times w \, Q/75]$$
$$= P /(w \, Q \, H/75)$$

12.2.12 Power produced by a reaction turbine: Work is done per kg of water, when it flows over the vanes. If we know the quantity of water flowing over the vanes per second, we can find out the amount of work done per second. The horsepower produced by the turbine is found by using the relation:

$$P = [\text{Work done/kg of water} \times \text{weight of water in kg flowing/sec}] / 75$$
$$= [\text{Total work done in kg m/sec}] / 75$$

12.3 Francis Turbine

The Francis Turbine is an inward flow reaction turbine, having radial discharge at outlet. It is mostly used for producing power under medium heads. The modern Francis turbine has a mixed flow (i.e. combination of radial and axial).

The height (or length) of the runner depends upon its specific speed. A Francis Turbine has a high specific speed with a longer runner than that having a lower specific speed. The runner of a Francis turbine is casted in one piece or made of separate steel plates welded together. The runners are made of cast iron for small outputs, cast steel for large outputs, and stainless steel, or other non-ferrous metal like bronze when water is chemically impure and there is a danger of corrosion. The blades of the runner are carefully finished.

All the relations for finding the various angles and other data which were used for inward flow reaction turbine will hold good for Francis turbine.

12.4 Kaplan Turbine

The Kaplan turbine is an axial flow reaction turbine in which the flow of water is parallel to the shaft. A Kaplan turbine is used when a large quantity of water is available at low heads. The runner of a Kaplan turbine resembles a propeller of a ship, for that reason a Kaplan turbine is also called a propeller turbine. The water from the scroll flows over the guide blades and over the vanes. The water exerts a force on the shaft of the turbine, which causes it to revolve.

The runner of a modern Kaplan turbine has the following major differences with that of Francis turbine:
(1) In a turbine runner, the water enters radially, whereas in a Kaplan turbine runner, the water strikes the blades axially.
(2) In a Francis turbine runner, the number of blades is generally between 16 and 24, whereas in a Kaplan turbine runner the number of blades is generally between 3 to 8. This reduces the frictional resistance.

The blades of a Kaplan turbine runner can be adjusted to control the passage area between the two blades.

The runner of a Kaplan turbine is known as boss, which is the extension of the shaft (as its end) shown in Figure (12-9):

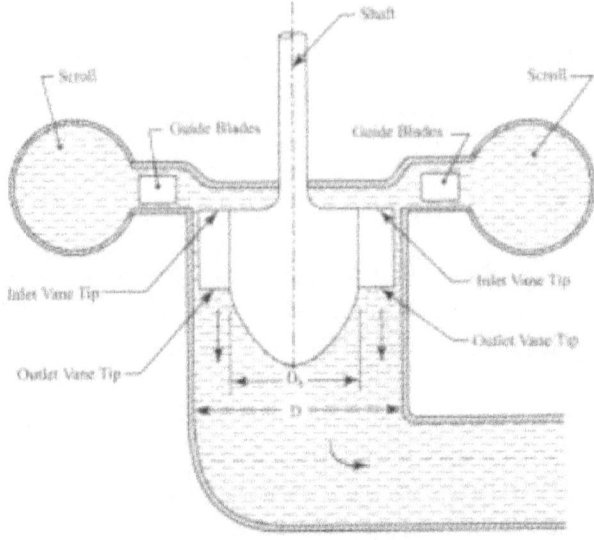

Fig (12-9) Kaplan turbine

Let,

D = diameter of turbine,

D_b= diameter of boss, and

V_f = velocity of flow at inlet.

The discharge through the turbine,

$$Q = V_f \times \pi/4 \; (D^2 - D_b^2)$$

All the notations and relations of the Kaplan turbine are the same as that of inward or outward flow reaction turbines.

12.5 Characteristics of Turbines

It will be more convenient when comparing the performance of turbines of different outputs and speeds working under different heads. To calculate outputs of the turbines when the head of water is reduced to unity, i.e. 1 metre, we always then study the following three characteristics:

1. Unit power
2. Unit speed, and
3. Unit discharge

12.5.1 Unit power: Is the power developed by a turbine, working under a head of 1 metre, is known as unit power (P_u).

Let,

H = head of water under which the turbine is working,

P = horse power developed by the turbine under a head of water H,

Q = discharge through the turbine, and

P_u= power developed by the turbine under a unit head.

The power developed by a turbine,

$$P = w \, Q \, H/75$$

As,

$$Q = \text{area of pipe x velocity of water in the pipe} = Av$$

And,

$$V = \sqrt{2gH}$$

Therefore,

$$P = [w \, (a \, \sqrt{2gH} \,) \, H] \, /75 \; \alpha \; H^{3/2} = \; P_u \; H^{3/2}$$

And,

$$P_u \; = \; P/H^{3/2}$$

12.5.2 Unit speed: The speed of a turbine working under a head of 1 metre is known as 'unit speed' (N_u)

Let,

H = head of water under which the turbine is working

v = tangential velocity of the runner,

N = speed of the turbine under a head of water H, and

N_u= speed of the turbine under a unit head.

The tangential velocity of runner is:

$$v \; \alpha \; \text{velocity of water} \; \alpha \; V$$

But,

$$V = (\sqrt{2gH})$$

Therefore,

$$v \; \alpha \; \sqrt{H}$$

And the tangential velocity of the runner is:

$$v = \pi \, D \, N/60$$

Or,

$$N = 60\, v\, / \pi\, D \quad \alpha\, v \quad \alpha\, \sqrt{H}$$
$$= N_u\, \sqrt{H} \qquad \text{(since, } v \ \alpha \ \sqrt{H})$$

Therefore,

$$N_u = N\, / \sqrt{H}$$

12.5.3 Unit discharge : The discharge of a turbine, working under a head of 1 metre, is known as unit discharge (Q_u).
Let
H = head of water under which the turbine is working,
Q = discharge through the turbine under a head of water H, and
Q_u = discharge through the turbine under a unit head.

The discharge through a turbine is:

$$Q = \text{area of pipe x velocity of water in the pipe} = Av$$

But,

$$V = \sqrt{2gH}$$

Therefore,

$$Q = a\, \sqrt{2gH} \quad \alpha\, \sqrt{H} \ = Q_u\sqrt{H}$$

And,

$$Q = Q\, / \sqrt{H}$$

12.6 Significance of Unit Power, Unit Speed and Unit Discharge

The conception of unit power, unit speed , and unit discharge is of much importance in the field of hydraulics. It helps in finding the behaviour of the turbine when put to work under different heads of water as discussed below:

12.6.1 Unit power: Let,
H = head of water under which the turbine is working,
P = power developed by the turbine under head of water H, and
P_1 = power developed by the turbine under another head of water H_1.

Since,

$$P \ \alpha \ H^{3/2}$$

Therefore,

$$P_1 \ \alpha \ H_1^{3/2}$$

Dividing one by the other:

$$P/P_1 = H^{3/2}/H_1^{3/2}$$

Therefore,

$$P_1 = P\,(H_1/H)^{3/2}$$

12.6.2 Unit speed: Let,

H = head of water under which the turbine is working,

N = speed of the runner under head of water H, and

N_1 = speed of the runner under another head of water H_1.

Since,

$$N \alpha \sqrt{H}$$

Therefore,

$$N_1 \alpha \sqrt{H_1}$$

Dividing one by the other:

$$N/N_1 = \sqrt{H} / \sqrt{H_1}$$

Therefore,

$$N_1 = N\,(H_1/H)^{1/2}$$

12.6.3 Unit discharge: Let,

H = head of water under which the turbine is working,

Q = discharge of the turbine under a head of water H, and

Q_1 = discharge of the turbine under water head H_1.

Since,

$$Q \alpha \sqrt{H}$$

Therefore,

$$Q_1 \alpha \sqrt{H_1}$$

Dividing one by the other:

$$Q/Q_1 = \sqrt{H} / \sqrt{H_1}$$

Therefore,

$$Q_1 = Q\,(H_1/H)^{1/2}$$

12.7 Specific Speed of a Turbine

After studying the behaviour of a turbine under unit conditions, the next step is to know the characteristics of an imaginary turbine identical (Identical means geometrically similar, having the same blade angles) with the actual turbine but reduced in size so as to develop a unit power under a unit head (i.e. horsepower under a head of 1 meter). Such a turbine condition is called the specific turbine and its speed is known as specific speed. Thus the specific speed of a turbine may be defined as the speed of an imaginary turbine, identical with the given turbine, which will develop a unit horsepower under a unit head.

Let,

N_S= specific speed of a turbine,

D = diameter of the turbine runner,

N = speed of the runner in rpm,

v = tangential velocity of the runner,

V = absolute velocity of the water.

The tangential velocity of the runner is:

$$v \, \alpha \, V$$

But,

$$V = \sqrt{2gH}$$

Therefore,

$$v \, \alpha \, \sqrt{2gH} \, \alpha \, \sqrt{H}$$

Also the tangential velocity of runner is:

$$v = \pi \, D \, N/ \, 60$$

Therefore,

$$DN \, \alpha \, v \, \alpha \, \sqrt{H} \qquad\qquad (\text{since, } v \, \alpha \, \sqrt{H})$$

And,

$$D \, \alpha \, \sqrt{H} \, / \, N \qquad\qquad (i)$$

Let,

Q = discharge through the turbine,

b = width of the turbine runner,

v_f = velocity of the flow, and

D = diameter of the turbine runner.

Since discharge,

$$Q = \pi \, D \, b \, v_f$$

But,

$$b \, \alpha \, D$$

And,

$$v_f \, \alpha \, \sqrt{2gH} \, \alpha \, \sqrt{H}$$

Therefore,

$$Q \, \alpha \, D^2 \sqrt{H}$$

Substituting the value of D^2 from equation (i) we get:

$$Q \, \alpha \, (\sqrt{H} / N)^2 \, \sqrt{H} \, \alpha \, H^{3/2}/N^2 \qquad\qquad \text{(ii)}$$

Let,

P = horsepower produced by the turbine.

Knowing that the power, $P = w \, Q \, H/75 \, \alpha \, Q \, H$

Substituting the value of Q from equation (ii):

$$P \, \alpha \, (H^{3/2}/ N^2 \, x \, H \, \alpha \, H^{5/2}/N^2$$

Or,

$$N^2 \alpha \, H^{5/2}/ P$$

Therefore,

$$N \, \alpha \, H^{5/2}/ \sqrt{P} = (N_S H^{5/4}) / \sqrt{P}$$

And,

$$N_S = N \, \sqrt{P} / H^{5/4}$$

12.7.1 Significance of specific speed: The significant feature of the specific speed of a turbine is that it is independent of the dimensions or size of both the actual and specific turbine. It is thus obvious that all the turbines, geometrically similar, working under the same heads and having same values of speed ratio and flow ratio will have the same specific speed.

In practice, the concept of specific speed is vital. The mere value of specific speed helps us in predicting the performance of a turbine as discussed below.

12.7.2 Selection of turbine: The selection of turbine is generally based on the following two factors:

1) Selection based on specific speed, and
2) Selection based on head of water.

The former (i.e. selection based on the specific speed) is the most scientific method and gives precise information, whereas the later (i.e. selection based on the head of water) is based on experience and observational factors.

12.7.3 Selection based on specific speed: The selection of type of turbines is made on the basis of acknowledged specific speed. The following table shows the type of turbine to be selected for the corresponding specific speed:

No	Specific Speed $[N_s = N \sqrt{P} / H^{5/4}]$	Type of Turbine
1	10 to 35	Pelton wheel with one nozzle
2	35 to 60	Pelton wheel with 2 or more nozzles
3	60 to 300	Francis turbine
4	300 to 1000	Kaplan turbine

12.7.4 Selection based on head of water: Following table shows the type of turbine used for the corresponding head of water:

No	Head of Water In metres	Type of Turbine
1	0 to 25	Kaplan or Francis *(preferably Kaplan)*
2	25 to 50	Kaplan or Francis *(preferably Francis)*
3	50 to 150	Francis
4	150 to 250	Francis or Pelton *(preferably Francis)*
5	250 to 300	Francis or Pelton *(preferably Pelton)*
6	Above 300	Pelton

12.8 Characteristic Curves of Turbines

A turbine is always designed and manufactured to work under a given set of conditions (or a limited range of conditions) such as discharge, head of water, speed, power generated, efficiency, etc...(at full speed or unit speed).

Since a turbine may have to be used under different conditions other than those for which it was designed, it is therefore essential that the exact behaviour of the turbine under varied conditions should be predetermined. This is represented graphically by means of curves, known as characteristic curves.

From the subject point of view the following characteristic curves are important:

(a) Characteristic curves for Pelton wheel, and
(b) Characteristic curves for Francis turbine.
   (or any other reaction turbine)

12.8.1 Characteristic curves for Pelton wheels:

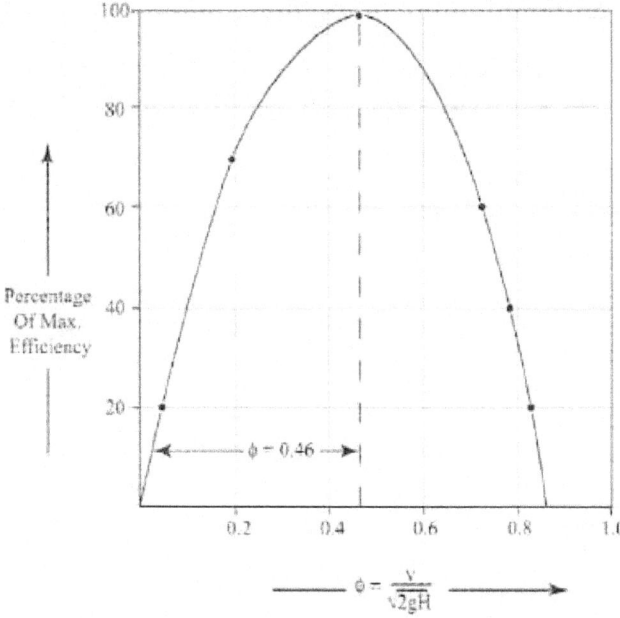

Fig (12-10) Speed ratio verses percentage of maximum efficiency

Figure (12-10) shows the performance of a Pelton wheel, under a uniform head and discharge. It is a parabolic curve, which shows that the efficiency increases from zero and beyond the value of $\varphi = 0.46$ the efficiency decreases.

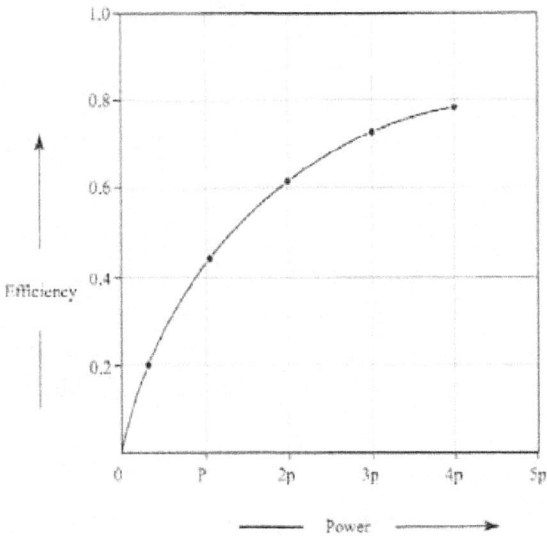

Fig (12-11) Power verses efficiency

Figure (12-12) shows the performance of a Pelton wheel under a constant head and speed. It is a parabolic curve which shows that the efficiency increases with the increase in power.

12.8.2 Characteristic curves of Francis turbines: In general the characteristic curves for Francis turbine (or any other reaction turbine) may be grouped under the following two heads:

(a) Characteristic curves for speed with varying heads, and
(b) Characteristic curves for speed under unit head.

12.8.3 Characteristic curves of Francis turbine for speed with varying heads:

(a) Speed verses discharge

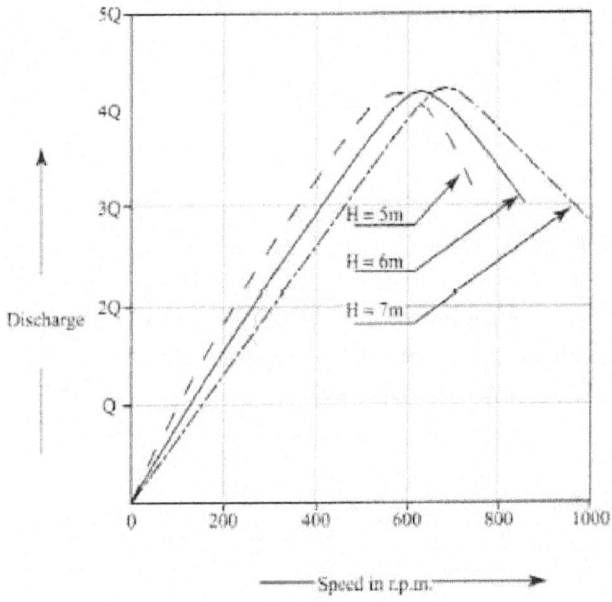

Fig (12-12) Speed verses discharge for Francis turbine

Figure (12-12) shows the performance of a Francis turbine (or any other reaction turbine) under variable heads, and constant discharge. It is a parabolic curve which indicates that for a given head the discharge increases with the speed from zero, and beyond a certain speed the discharge decreases.

(b) Speed verses power

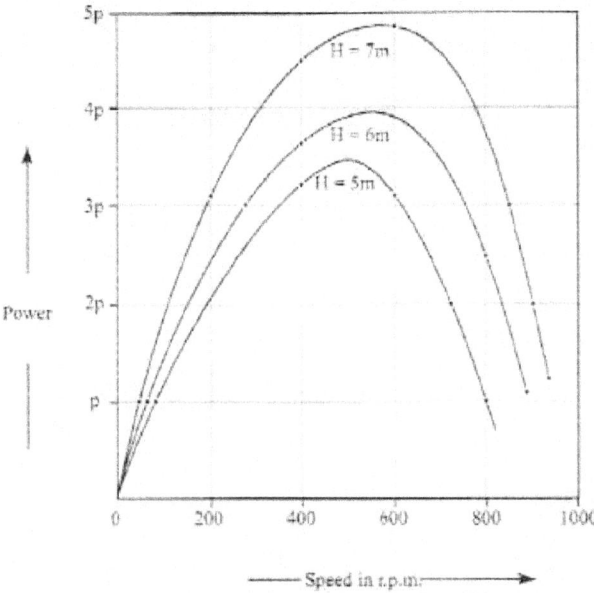

Fig (12-13) Speed verses power for Francis turbine

Figure (12-13) shows the performance of a Francis turbine (or any other reaction turbine) under variable head and constant discharge. It is a parabolic curve which indicate that for a given head the power increases with the speed from zero, and beyond a certain speed the power decreases.

(c) Speed verses efficiency

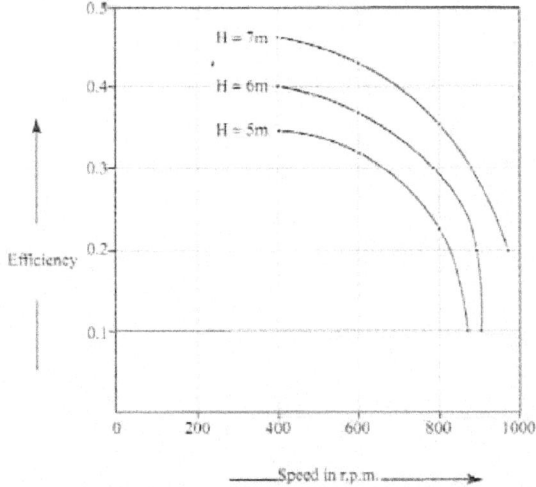

Fig (12-14) Speed verses efficiency for Francis turbine

Figure (12-14) shows the performance of a Francis turbine (or any other reaction turbine) under variable heads and constant discharge. It is a parabolic curve, which indicated that for a given head the efficiency decreases with the increase in speed.

12.8.4 Characteristic curves of Francis turbine for speed under unit head:

(a) Speed under unit head verses discharge

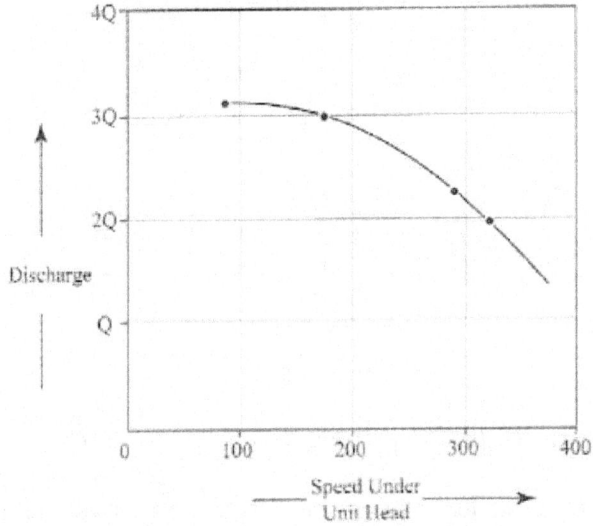

Fig (12-15) Speed under unit head verses discharge for Francis turbine

Figure (12-15) shows the performance of a reaction turbine. It is a parabolic curve which indicates that the discharge decreases with the speed under unit head.

(b) Speed under unit head verses power

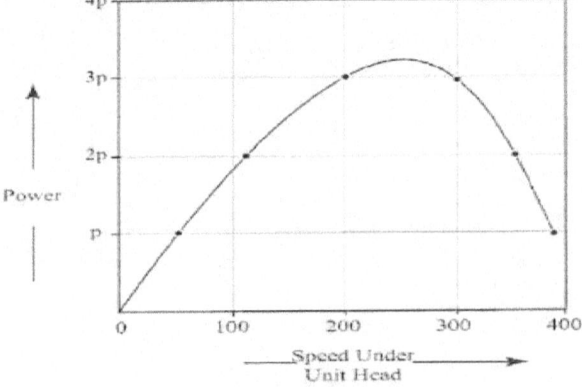

Fig (12-16) Speed under unit head verses power for Francis turbine

Figure (12-16) shows the performance of a reaction turbine. It is a parabolic curve which indicates that the power increases with the speed under unit head, and beyond a certain speed the power decreases.

(c) Speed under unit head verses efficiency

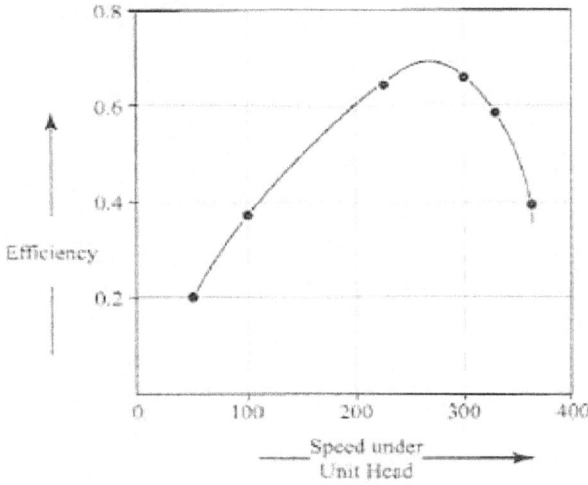

Fig (12-17) Speed under unit head verses efficiency for Francis turbine

Figure (12-17) shows the performance of a reaction turbine. It is a parabolic curve which indicates that the efficiency increases with the speed under unit head, and beyond a certain speed the efficiency decreases.

Check Your Knowledge
1) What is meant by an impulsive turbine?
2) State the difference between an impulse turbine and a reaction turbine.
3) Describe with the help of simple sketches the action of the impulse turbine.
4) Drive an equation for the hydraulic efficiency of a Pelton wheel.
5) What do you understand from the term turbine? Discuss briefly the uses of turbines.
6) Distinguish clearly between radial flow and axial flow turbines.

7) Write the difference between the inward flow turbines and outward flow turbines.

8) Derive an equation for the horse power developed by a reaction turbine.

9) Define the term 'unit power', 'unit speed', and 'unit discharge' with reference to a hydraulic turbine.

10) Derive expressions for unit power, unit speed and unit discharge for a turbine.

11) What factors decide whether a Kaplan, Francis or a Pelton type turbine would be used in a hydro-electric project?

Exercise

1) A Pelton wheel working under a head of 380 metres develops 16250 horsepower at the rate of 750 rpm. Find the diameter of the wheel and number of jets, if the overall efficiency of the wheel is 86%. [**Ans.** 1m; 4]

2) A twin jet Pelton wheel is required to work under a head of 50 metres and to produce 120 horsepower running at 300 rpm. Assuming overall efficiency of 90% and coefficient of velocity as 0.86, find:

(a) diameter of jet,

(b) width of the buckets,

(c) depth of the buckets, and

(e) number of buckets.

[**Ans.** 6.5 cm; 32.5 cm; 7.8 cm; 22]

3) An inward flow reaction turbine, having external diameter of 1 metre is running at a speed of 180 rpm. The guide blade angle is 15°. If the velocity of flow at inlet is 3 metres/sec, find:

(a) peripheral velocity at inlet,

(b) velocity of whirl at inlet,

(c) absolute velocity of water at inlet, and

(d) vane angle at inlet.

[**Ans.** 9.45 m/sec; 11.3 m/sec; 11.68 m/sec; 59° 30′]

4) An inward flow reaction turbine, having external diameter of 50 cm is running at 200 rpm. If the discharge is radial at outlet, find the peripheral velocity at outlet and the vane angle at outlet. [**Ans.** 7.41 m/sec; 32° 28′]

5) The external and internal diameters of an inward flow reaction turbine are 2 metres and 1 metre respectively. The turbine is running at a speed of 192 rpm. The guide blade angle is 10° and velocity of flow at inlet and outlet is 5 metres/ sec. Draw the velocity triangles at inlet and outlet and find:

(a) vane angle at inlet and outlet, and

(b) Absolute velocity of water leaving the guide vanes. [**Ans.** 20° 54′; 25°30′; 28.8 m/sec]

6) A Francis turbine having outer diameter 90 cm is running at 200 rpm. The head of water on the turbine is 9.5 metres. The velocity of flow through the runner is constant at 3

metres/sec. If the inlet tips of the vanes are radial and the width of the runner at inlet is 15 cm, find:

(a) work done / kg of water,

(b) hydraulic efficiency of the turbine, and

(c) HP produced by the turbine.

[**Ans.** 9.04 kg-m; 95.19%; 153.3 HP]

7) A Kaplan turbine working under a head of 5.5 metres develops 10000 h.p. The speed ratio and flow ratio are 2.10 and 0.71 respectively. If the loss diameter is 1/3 that of the runner and the overall efficiency is 85%, find:

(a) Diameter of runner, and

(b) Speed of the turbine.

[**Ans.** 5.58 m; 75 rpm]

8) A turbine working under a head of 23.4 metres develops 28,000 horsepower at 250 rpm. Determine unit power and unit speed. [**Ans.** 25.2 HP; 51.6 rpm]

9) A turbine develops 500 horsepower under a head of 100 metres at 200 rpm. What would be its normal speed and output under a head of 81 metres? [**Ans.** 180 rpm; 364.5 hp]

10) Find the specific speed of a turbine developing 10000 horsepower under a head of 27 metres at 120 rpm [**Ans.** 195 rpm]

11) Find the type of turbine developing 1700 hp under a head of 150 m at 375 rpm to discharge 200 litres of water/sec. [**Ans.** N_S = 29.4; Pelton wheel with 1 nozzle]

13. Centrifugal Pumps

13.1 Introduction

Since olden times man has been curious to find convenient ways to lift water to higher levels for water supply and irrigation purposes. It is believed that the idea of lifting water by centrifugal force was first put forward by Lorand D Vinci at the end of the 15th century. This idea was under investigation by French scientists who designed centrifugal pumps with impeller and blades. At that time reciprocating pumps were very popular. The continuous advancement of this pump has brought it to a high degree of perfection.

13.2 Type of Pumps

Though there were many types of pumps the following are important from this subjects point of view:

1. Centrifugal pumps, and
2. Reciprocating pumps.

This chapter will discuss centrifugal pumps and their characteristics.

13.2.1 Centrifugal pump: A pump, in general, may defined as a machine which when driven from some external source lifts water or some other liquid from a lower level to a higher level. In other words, a pump may also be defined as a machine which converts mechanical energy into pressure energy. The pump which uses water or a liquid from a lower level to a higher level by centrifugal force is known as a centrifugal pump.

The action of a centrifugal pump is that of a reversed reaction turbine. In a reaction turbine, the water at high pressure is allowed to enter the casing which gives out mechanical energy at its shaft, whereas in a pump the mechanical energy is fed into the shaft and water enters the impeller (attached to the rotating shaft) which increases the pressure energy of the outgoing fluid. The water enters the impeller radially.

The work done or the power required by the pump may be found out by drawing the inlet and outlet triangles, as shown in Figure (13-1):

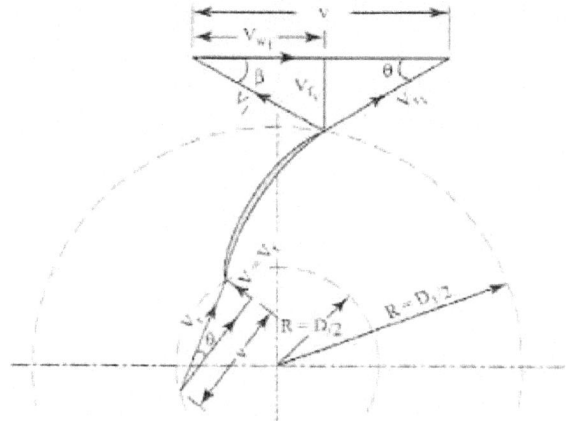

Fig (13-1) Inlet and outlet triangles for a centrifugal pump

Let,

V = water absolute velocity at entrance,

V_1 = water absolute velocity at exit,

D = diameter of impeller at inlet (inner diameter)

D_1 = diameter of impeller at outlet (outer diameter)

N = speed of impeller in rpm,

υ = tangential velocity of impeller at inlet (also known as peripheral velocity at outlet),

υ_1 = tangential velocity of impeller at outlet (also known as peripheral velocity at outlet),

V_r = relative velocity of water to the wheel at inlet,

V_{r1} = relative velocity of water to the wheel at outlet,

V_f = velocity of flow at inlet,

V_{f1} = velocity of flow at outlct,

Θ = vane angle at inlet

β = angle at which the water leaves the impeller, and

ϕ = vane angle at outlet.

Since the water enters the impeller radially therefore the velocity of whirl at inlet (V_w = 0), and:

$$\text{Work done per kg of water} = V_{w1} \, \upsilon_1/g$$

13.2.2 Types of casing for an impeller in a centrifugal pump: A centrifugal pump consists of an impeller, similar to that of a turbine, to which curved vanes are fitted. The impeller is closed in a watertight casing, having a delivery pipe in one of its sides. The casing for a centrifugal pump is so designed that the kinetic energy of the water is converted into pressure energy. Following are the two types of casing or chambers:

1. Volute casing, and
2. Volute casing with guide blades.

13.2.3 Volute casing (spiral casing):

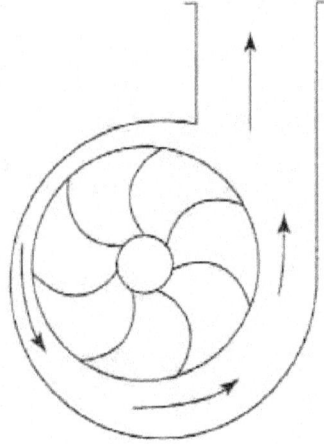

Fig (13-2) Volute of a centrifugal pump

In a volute chamber, the impeller is surrounded by a spiral casing as shown in Figure (13-2). Such a casing provides a gradual increase in the area of flow; thus decreasing the velocity of water, and correspondingly increasing the pressure. A considerable loss takes place due to the formation of eddies in this type of casing.

13.2.4 Volute casing with guide blades:

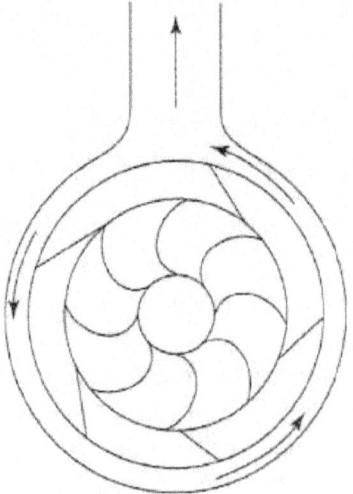

Fig (13-3) Volute with guide vanes

In this type of casing guide blades are surrounding the impeller as shown in Figure (13-3). These guide blades are arranged at such an angle so that the water enters without shock forming a passage of increasing area up to the delivery pipe.

13.3 Manometric Head

It is an important term and may be defined in any of the following four ways:

1) The manometric head is the actual head against which the pump has to work.

2) Manometric head,

$$H_m = H_s + H_{fs} + H_d + H_{fd} + V_d^2/2g$$

Where,

H_s = suction lift,

H_{fs} = loss of head in suction pipe due to friction,

H_d = delivery lift,

H_{fd} = loss of head in delivery pipe due to friction, and

V_d = velocity of water in the delivery pipe.

3) Manometric head,

$$H_m = \text{work done/kg of water - loss within the impeller}$$
$$= V_{w1}\, \upsilon_1/g - \text{impeller losses}$$

4) Manometric head,

$$H_m = \text{energy/kg at outlet of impeller - energy/kg at inlet of impeller}$$

13.4 Efficiencies of a Centrifugal Pump

A centrifugal pump has the following three types of efficiencies:

1) Hydraulic efficiency,
2) Mechanical efficiency, and
3) Overall efficiency.

13.4.1 Hydraulic efficiency: It is the ratio of manometric head, to the energy of the impeller/kg of water. Mathematically, the hydraulic efficiency is defined as:

$$\eta_h = H_m / (V_{w1}\, \upsilon_1/g)$$

13.4.2 Mechanical efficiency: It is the ratio of energy available at the impeller to the energy given to impeller by prime power.

13.4.3 Overall efficiency: It is the ratio of the actual work done by the pump to the energy supplied to the pump by a prime mover.

13.4.4 Discharge of a centrifugal pump: The discharge of a centrifugal pump is calculated in a similar manner as that of a reaction turbine i.e.

Let,

D = diameter of impeller at inlet,

V_f = velocity of flow at inlet,

b = width of impeller at inlet, and

$D_1 \, b_1 \, V_{f1}$ = corresponding values at the outlet.

Then the discharge,

$$Q = \pi \, D \, b \, V_f = \pi \, D \, b \, V_{f1}$$

13.5 Increase in the Water Pressure within an Impeller of a Centrifugal Pump

A pump converts mechanical energy into pressure energy. This pressure energy is given by the impeller to the fluid flowing through it.

Consider a centrifugal pump as shown in Figure (13-4):

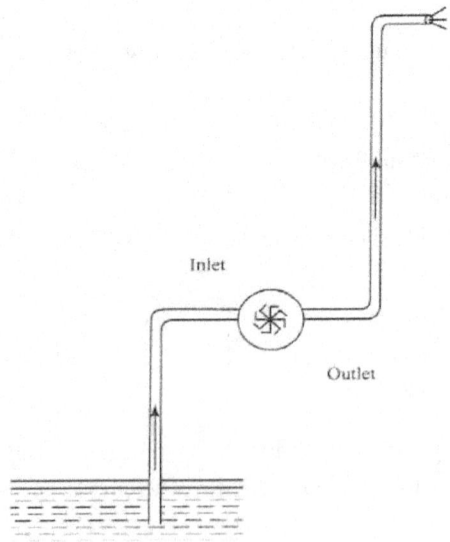

Fig (13-4) Centrifugal pump during operation

Applying Bernoulli's equation to the inlet outlet of the impeller we get:

Energy at outlet = Energy inlet + work done by the impeller

i.e. $p_1/w + V_1^2/2g = p/w + V^2/2g + V_{w1}\, \upsilon_1/g$ (taking $Z_1 = Z$)

$p_1/w - p/w = V^2/2g - V_1^2/2g + V_{w1}\, \upsilon_1/g$

$p_1/w - p/w$ represents the increase in fluid pressure.

13.6 Minimum Starting Speed for Centrifugal Pump

A centrifugal pump will start delivering liquid only when the head developed by it is equal to the manometric head. At start time the liquid velocity is zero, therefore,

The pressure head caused by the centrifugal force = $V_1^2/2g - V^2/2g = (V_1^2 - V^2)\, /\, 2g$
This pressure head will give the required manometric head, therefore,

$$(V_1^2 - V^2)\, /\, 2g = H_m$$
$$= \eta_h\, (V_{w1}\, \upsilon_1/g) \qquad [\text{since, } \eta_h = H_m/\, (V_{w1}\, \upsilon_1/g)]$$

13.7 Multistage Centrifugal Pumps

We have seen that the head developed by a centrifugal pump is proportional to its diameter and speed of the impeller.

Since there is a limitation for the diameter and speed of the impeller, therefore the head developed by a centrifugal pump is limited to 50 meters. In some special pumps higher heads up to 100 metres may also be developed. For still larger heads, it will be necessary to put two or more pumps in series. Liquid from one pump is brought to the inlet of another pump, which further increases the head developed.

This type of arrangement is also possible if we provide two or more impellers (instead of two or three pumps) keying them to the same shaft and putting them in the same casing. Such a pump is called a Multi-stage pump.

13.8 Specific Speed of a Centrifugal Pump

The specific speed of a centrifugal pump is defined as the speed of an imaginary pump, identical to the given pump, which will discharge one litre of water while it is being raised through a head of 1 metre.

Let,

N_s = specific speed of the pump,

D_1 = diameter of the impeller at outlet,

N = speed of the impeller in rpm,

υ_1 = tangential velocity of impeller at outlet, and

H = lift of the pump.

The tangential velocity of the impeller,

$$\upsilon_1 \alpha \sqrt{H}$$

And the tangential velocity of the impeller,

$$V = \pi D_1 N / 60$$

Therefore,

$$D_1 N \alpha \upsilon \alpha \sqrt{H}$$

Therefore,

$$D_1 \alpha \sqrt{H} / N$$

Let,

Q = discharge of the pump,

b_1 = width of the impeller at outlet, and

V_{f1} = velocity of flow at outlet.

Since, $Q = \pi D_1 b_1 V_{f1}$

But,

$$b_1 \alpha D_1$$

And,

$$V_{f1} \alpha \sqrt{H} \qquad \text{(since, } D_1 \alpha \sqrt{H} / N\text{)}$$
$$Q \alpha D_1^2 \sqrt{H} \alpha (\sqrt{H} / N)^2 \sqrt{H} \alpha H^{3/2} / N^2$$

Therefore,

$$N^2 \alpha H^{3/2} / Q$$

Or,

$$N \alpha H^{3/2} / \sqrt{Q} = N_s H^{3/4} / \sqrt{Q}$$

Or,

$$N_s = N \sqrt{Q} / H^{3/4}$$

13.8.1 Selection of centrifugal pumps based on specific speed: The specific speed of a centrifugal pump, like that of a turbine, helps us in selecting the type of centrifugal pump. The following table gives the type of centrifugal pump for the corresponding specific speed:

No	Specific Speed	Type of Centrifugal Pump
1.	10 to 30	Slow speed pump, with radial flow at outlet
2.	30 to 50	Medium speed pump, with radial flow at outlet
3.	50 to 80	High speed pump, with radial flow at outlet
4.	80 to 160	High speed pump, with mixed flow at outlet
5.	160 to 500	High speed pump, with axial flow at outlet

13.9 Priming of a Centrifugal Pump

The pressure developed by the impeller of a centrifugal pump is proportional to the density of the fluid in the impeller. It is thus obvious that an impeller running with air will produce only a negligible pressure, which may not be able to transport the fluid from its source through the suction pipe. To avoid this, the pump is first primed i.e. filled up with liquid.

To do so, the suction pipe and the impeller is completely filled with fluid. The delivery valve is closed and the pump is started. The rotating impeller pushes the fluid in the delivery pipe, opens the delivery valve and transports the liquid through the suction pipe.

13.10 Characteristic Curves of Centrifugal Pumps

A centrifugal pump, like a turbine, is designed and manufactured to work under a given set of conditions (or a limited range of conditions) such as discharge, speed, power required, head of water, efficiency etc. In addition, a pump may have to work under conditions other than those for which it has been designed for.

As a result, the exact behaviour of the pump under varied conditions should be predetermined. This is represented graphically by means of curves, known as 'characteristic curves'. Tough there are many types of characteristic curves the following are most important from the subject point of view:
(1) Characteristic curves for discharge with varying speeds, and
(2) Characteristic curves for speed.

13.10.1 Characteristic curves of centrifugal pumps for discharge with varying speeds:

a) Discharge versus head:

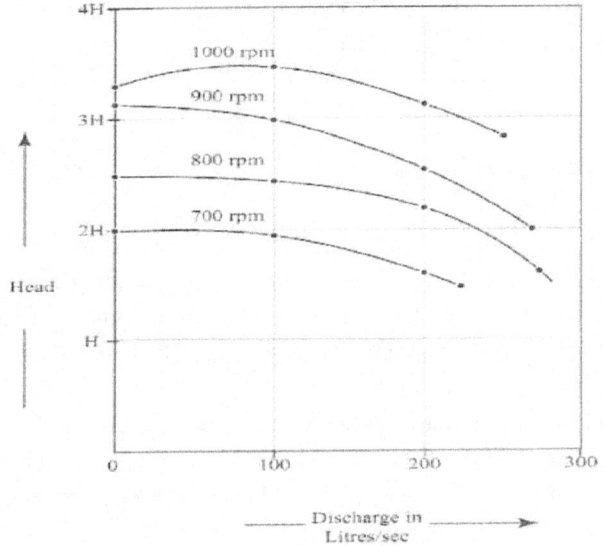

Fig (13-5) Q-H curve

Figure (13-5) shows the performance of a centrifugal pump under variable rotational speeds. It is a parabolic curve that for a given rotational speed the manometric head slightly increases with the discharge and beyond a certain quantity of discharge the head decreases.

b) Discharge versus power:

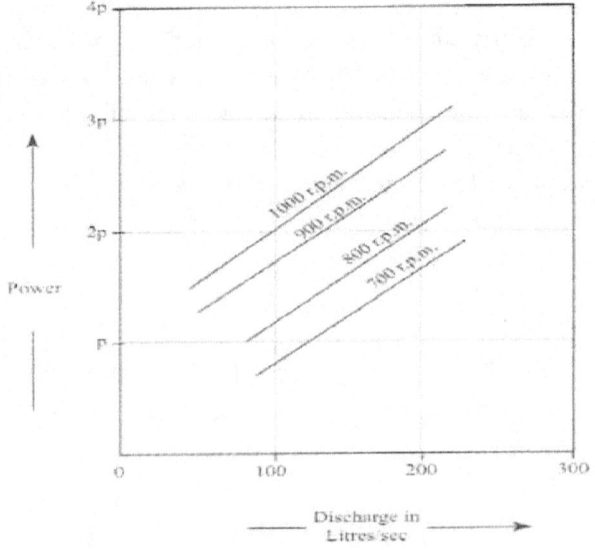

Fig (13-6) Q-P curve

Figure (13-6) shows the performance of a centrifugal pump under variable rotational speeds. It is almost a straight line curve, which shows that for a given rotational speed the power increases gradually with discharge.

c) Discharge versus efficiency:

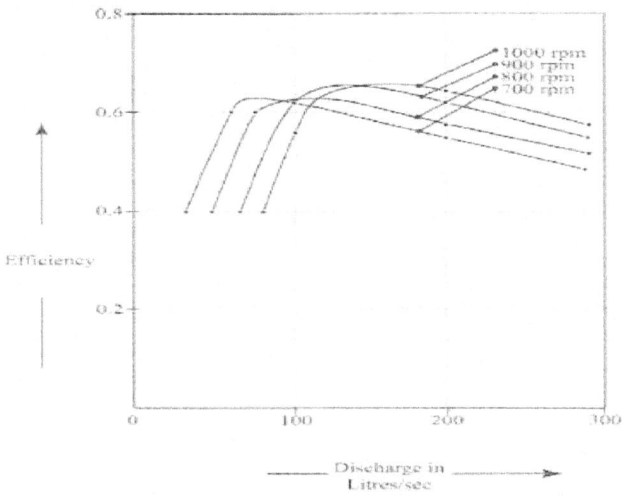

Fig (13-7) Q-η curve

Figure (13-7) shows the performance of a centrifugal pump under variable rotational speeds. This is a parabolic curve indicating that for a given rotating speed the efficiency increases with discharge, and beyond a certain discharge the efficiency start decreasing.

13.10.2 Characteristic curve of centrifugal pump for speed:

a) Speed verses discharge:

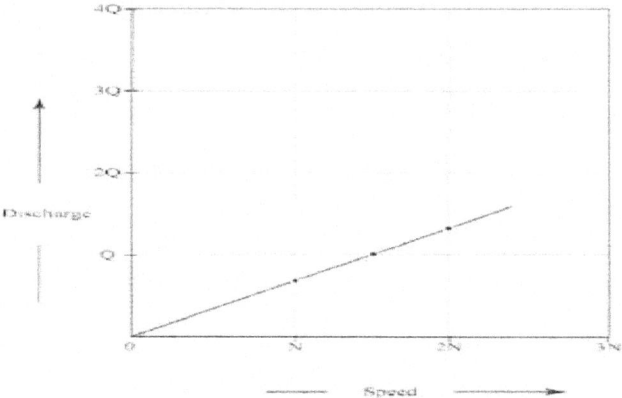

Fig (13-8) N-Q curve

Figure (13-8) shows the performance of a centrifugal pump under constant head. This is a straight line which indicates that the discharge increases with the speed.

b) Speed versus power:

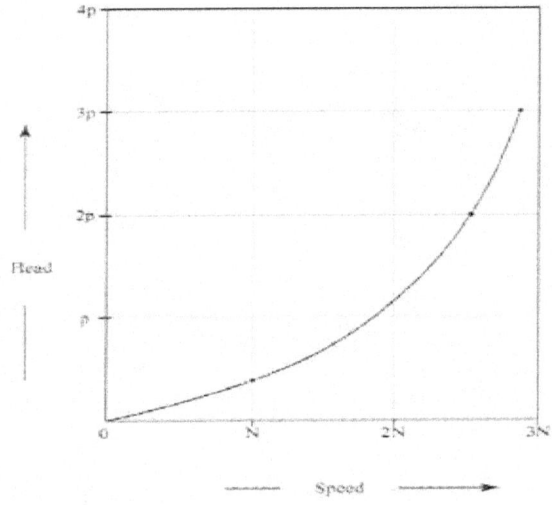

Fig (13-9) N-P curve

Figure (13-9) shows the performance of a centrifugal pump under a constant head and discharge. This is a parabolic curve which indicates that the power increases with the speed.

c) Speed versus head:

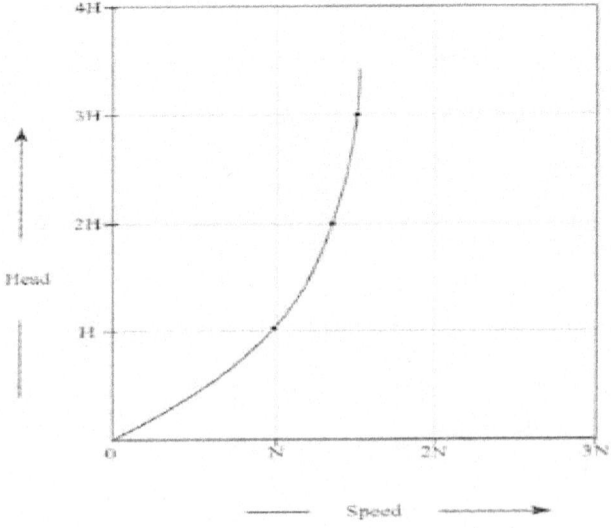

Fig (13-10) N-H curve

Figure (13-10) shows the performance of a centrifugal pump under a constant discharge. This is a parabolic curve, which indicates the head increases with the speed.

13.11 Cavitation in Pumps

As the impeller blades move through a fluid, low-pressure areas are formed due to the fluids acceleration moving past the blades. The faster the blade moves, the lower the pressure which at a stage vaporises the fluid and forming small bubbles of gas. As the bubbles collapse they cause very strong local shock waves, which may be audible and may even damage the blades.

Cavitation creates problems during the operation of all three types of centrifugal pumps (radial, mixed, and axial flow pumps) whenever high discharge, high rotational speed, or low head is encountered. Pumps with low specific speed are more susceptible to cavitation as compared to high specific speed pumps.

Cavitation in pumps occurs in two different forms:

1- Suction cavitation occurs when the pump suction is under a low-pressure/high-vacuum condition as the liquid turns into vapour at the eye of the pump impeller. This vapour is carried over to the discharge side of the pump, where it no longer sees vacuum and is compressed back into a liquid by the discharge pressure. This imploding action occurs violently and attacks the face of the impeller.
An impeller that has been operating under a suction cavitation condition can have large shanks of material removed from its face or very small bits of material removed, causing the impeller to look like a sponge. Both cases will cause premature failure of the pump, often due to bearing failure. Suction cavitation is often identified by a sound like gravel or marbles in the pump casing.
In automotive applications, a clogged filter in a hydraulic system can cause suction cavitation, making a noise that rises and falls in synchronic rhythm with the engine RPM.

2- Discharge cavitation occurs when the pump discharge pressure is extremely high; normally occurring in a pump that is running at less than 10% of its best efficiency point. The high pressure causes the majority of the fluid to circulate inside the pump instead of being allowed to flow out the discharge. As the liquid flows around the impeller, it must pass through the small clearance between the impeller and the pump casing at extremely high velocity. This velocity causes a vacuum to develop at the casing wall, which turns the liquid into vapour. A pump that has been operating under these conditions shows premature wear of the impeller vane tips and the pump casing. In addition, due to the high pressure conditions, premature failure of the pump's mechanical seal and bearings can be expected. This can break the impeller shaft

Check Your Knowledge

1) What is a centrifugal pump? On what principle does it work?

2) What are the different types of pumps? Explain the working principles of a centrifugal pump with sketches.

3) Name the different types of casings of the impeller of a centrifugal pump.

4) Explain the function of spiral casing for a centrifugal pump.

5) Name the different types of efficiencies of a centrifugal pump and differentiate between overall efficiency and manometric efficiency.

6) Obtain an equation for the increase in water pressure while flowing through the impeller of a centrifugal pump.

7) What do you understand by the term 'multistage pump'? Explain clearly the difference between a single stage and a multistage centrifugal pump.

8) Write an expression for the specific speed of a centrifugal pump.

9) What is priming of a centrifugal pump? Explain clearly why priming is essential before starting a centrifugal pump?

10) Draw the following characteristic curves of a centrifugal pump:

      (i) Head, power and efficiency verses discharge with constant speed.

      (ii) Head, power and efficiency verses discharge with variable speed.

11) Explain the occurrence of cavitation in pumps.

14. Hydraulic Devices

14.1 Introduction

Various types of machines were discussed in which liquid (water) moves from one point to another. In some of the machines (i.e. turbines) the potential or kinetic energy of liquid is converted into mechanical energy. But in other machines (i.e. pumps) mechanical energy is converted into potential or kinetic energy.

In this chapter hydraulic machines is discussed in which the liquid (water or oil) acts as a medium of transmission for power or pressure, based on the principles of hydrostatics and hydraulics.

Though there are numerous devices based on the principle of hydrostatics and hydraulics, the following are important from the subject point of view:

(1) The hydraulic press,
(2) The hydraulic accumulator,
(3) The hydraulic intensifier,
(4) The hydraulic ram,
(5) The hydraulic crane, and
(6) The hydraulic lift.

14.1.1 The hydraulic press: It is a device by which one can lift a larger load by the application of a comparatively smaller force.

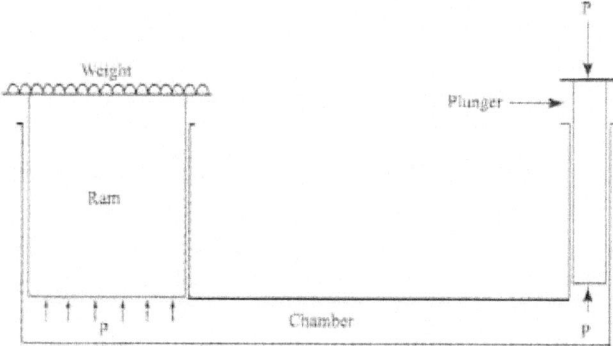

Fig (14-1) Hydraulic Press

A hydraulic press, in its simplest form, consists of two cylinders, one larger and the other smaller, connected to a chamber containing liquid. The larger cylinder contains a ram and a smaller cylinder a plunger, as shown in Figure (14-1).

A smaller force P acts on the plunger, in the downward direction, which presses the liquid. This is transmitted equally in all directions and raises the ram. The heavier load, placed on the ram, is then lifted up.

Let,

A = area of the ram,

a = area of the plunger,

p = intensity of pressure,

P = force applied on the plunger, and

W = weight lifted by the ram.

Since the intensity of pressure in the chamber is the same in all the directions, therefore:

$$P = p\,a$$

Or,

$$P = P/a \tag{i}$$

And,

$$W = p\,A$$

Or,

$$P = W/A \tag{ii}$$

Equating (i) and (ii):

$$P/a = W/A$$

Or,

$$W = P \times A/a$$

Or,

$$P = W \times a/A$$

If losses are taken into account, then the efficiency of the hydraulic press is:

$$\eta = (W/A) / (P/a) = W/P \times a/A$$

The ratio of load lifted to the effort applied (i.e. W/P) is known as the mechanical advantage.

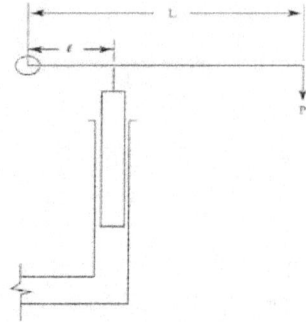

Fig (14-2) Plunger with a lever

The mechanical advantage of a hydraulic press may be increased by applying a force on the plunger using a lever as shown in Figure (14-2). The ration L/l is known as the leverage of the press.

14.1.2 The hydraulic accumulator: This is a device made to store pressure energy, which is ready to supply to machines at a later stage.

A hydraulic accumulator, in its simplest form, consists of a vertical cylinder containing a sliding ram as shown in Figure (14-3):

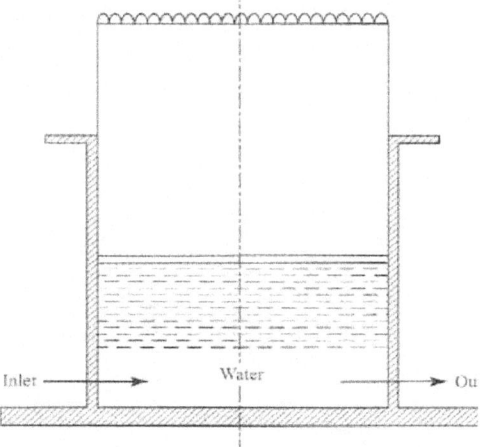

Fig (14-3) Hydraulic ram

The sliding ram is loaded with weights. When the water, under pressure, enters the accumulator cylinder through the inlet value, it lifts the loaded ram until the cylinder is full of water.

At this stage, the accumulator has stored its maximum amount of energy, which is also known as the capacity of its accumulator.
Let,
A = area of the cylinder,
H = lift of the cylinder, and
p = intensity of pressure of the water supplied by the accumulator,

Then,

$$\text{The capacity of accumulator} = p\,A\,H \qquad \text{kg-m}$$

14.1.3 The hydraulic intensifier: This is a device to increase the intensity of pressure of water by means of energy available from a larger quantity of water at a low pressure. A hydraulic intensifier, in its simplest form, consists of a fixed ram through which the water, under high pressure, flows to the machine.

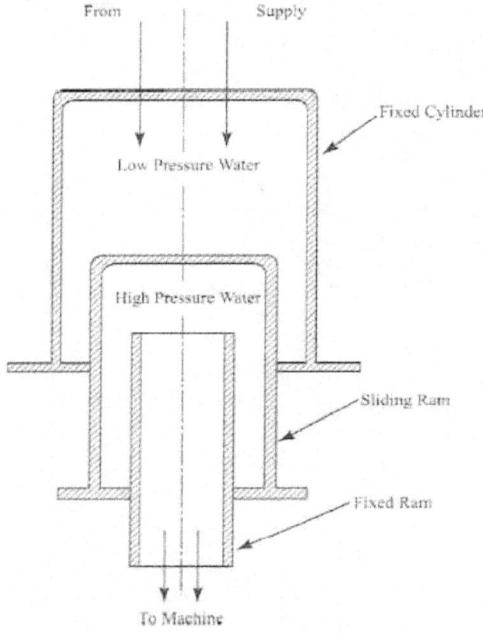

Fig (14-4) Hydraulic intensifier

A hollow sliding ram is mounted externally on this fixed ram, which contains water under high pressure as shown in Figure (14-4). The sliding ram is encased in a fixed cylinder which contains water under a low pressure from the supply as shown.

The water, under a low pressure, presses the sliding ram on the top forcing it downwards on to the fixed ram. This downward movement of the sliding ram increases the intensity of pressure of the water in the sliding ram.

Let,

A = external area of the sliding ram,

a = area of sliding ram end,

p = intensity of pressure (of low pressure) in the fixed cylinder, and

P = intensity of pressure (of high pressure) in the sliding ram.

Therefore,

$$\text{Total upward force} = \text{Area x Intensity of pressure}$$
$$= a \times P \qquad\qquad (i)$$

Similarly,

$$\text{Total upward force} = p \times A \qquad\qquad (ii)$$

Equating (i) and (ii),

$$P \times a = p \times A$$

Therefore,

$$P = A/a \times p$$

14.1.4 The hydraulic ram: This is an automatic machine which can lift a small quantity of water to a greater height, when a large quantity of water is available at a smaller height.

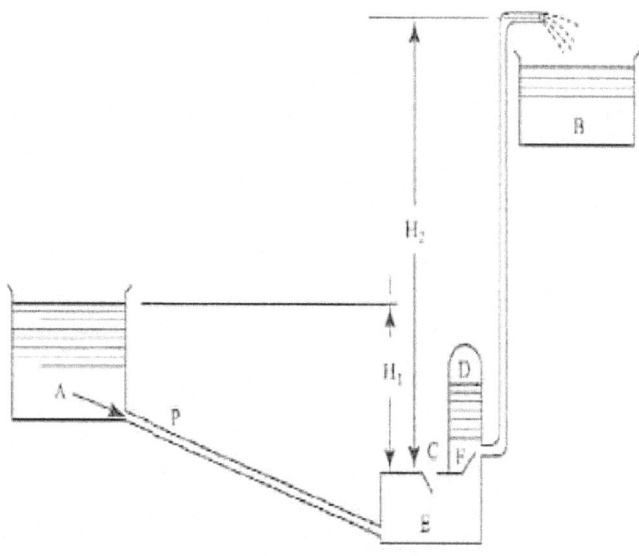

Fig (14-5) Hydraulic ram

Figure (14-5) shows a diagrammatic view of a hydraulic ram, in which water available from source A at a height H_1. By means of the hydraulic ram, a small quantity of this water is raised through a height H_2 into the tank B.

The water starts flowing from tank A to the chamber E through the pipe P. The waste-valve C is open and the water flows out. As the speed of the water in the pipe P increases, the dynamic pressure on the waste valve C increases until it is greater than the weight of the valve lid, which will suddenly close the valve. This sudden closure of the valve brings the water in the pipe and chamber suddenly to rest, which will increase the pressure in chamber E. This increase in pressure lifts the valve F and some water flows into the air vessel D, which will compress the air in the air-vessel causing its pressure to increase. Thus increased air pressure in the air vessel D forces the water upward into tank B.

When the momentum of the water in chamber E is destroyed, the waste-valve C opens and F closes, which causes the flow of water from the tank A to recommence.

Let,

W = weight of the water flowing from tank A into the Chamber E,

w = weight of the water flowing from the chamber E into the tank B,

H_1 = height of water in the tank A, above the Chamber E, and

H_2 = height of water in the tank B, above the Chamber E.

$$\text{Energy supplied by the tank A} = W\,H_1 \qquad\qquad \text{(i)}$$
$$\text{Energy supplied by the tank B} = w\,H_2 \qquad\qquad \text{(ii)}$$

Equating (i) and (ii):

$$W\,H_1 = w\,H_2$$
$$w = (H_1/H_1) \times W$$

If losses are taken into account, then efficiency of the ram (known as D'Aubuisson's efficiency) is:

$$\eta = (w\,H_2) / (\,W\,H_1)$$

There is another relation for the efficiency of the ram (known as Rankine's efficiency). In this relation it is assumed that the water was initially at a height H_1 and is only lifted through a height equal to $(H_2 - H_1)$. Thus the Rankine's formula for the efficiency of a ram,

$$\eta = [w\,(H_2 - H_1)] / (W\,H_1)$$

Note: If instead of the weight of water in the two tanks, the two discharges (i.e. from the tank A to the chamber, as Q and from the chamber E to the tank B, as q) are given for the ram, then the efficiency of the ram will be given by the relation:

$$\eta = (q\,H_2) / (Q\,H_1)$$

And the corresponding Rankine's efficiency will be given by the relation:

$$\eta = [q\,(H_2 - H_1)] / (Q\,H_1)$$

14.1.5 Hydraulic crane: This is a device for raising or transferring heavy loads (up to 250 tonnes) and is widely used on docks, sidings, warehouses or workshops.

A hydraulic crane consists of a vertical post, tie and jib (i.e. basic requirements of a crane) having guide pulleys. Near the foot of the vertical crane post is provided a jigger. The jigger consists of a fixed cylinder having a pulley block and containing a sliding ram. One end of the ram is in contact with the water and the other carries a pulley block.

A chain or wire rope, one end of which is fixed, is taken round all the pulleys of the two blocks, through the vertical post and finally over the guide pulleys. A hook is attached to the other end of the rope for handling the load as shown in Figure (14-6). There is a pipe connection for supplying water under a high pressure to the fixed cylinder of the jigger as shown.

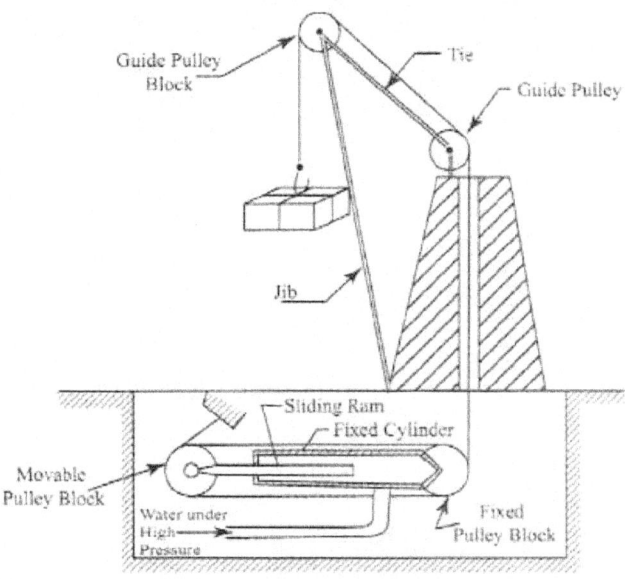

Fig (14-6) Hydraulic crane

The load to be lifted is suspended on the free end of the wire rope. The water, under a high pressure is admitted into the cylinder of the jigger. This water forces the sliding ram to move towards the left. This outward movement of the sliding ram, makes the pulley block move outwards. Due to the increased distance between the two pulley blocks the wire rope is pulled and the load is lifted up.

If the load is required to be shifted, the vertical post of the crane can be rotated through the desired horizontal angle.

14.1.6 The hydraulic lift: This device is designed to lift or bring down loads or passengers from one floor to another in a multi-storeyed building. Following are the two types of hydraulic lifts:

(a) Direct acting hydraulic lift, and
(b) Suspended hydraulic lift.

(a) Direct acting hydraulic lift:

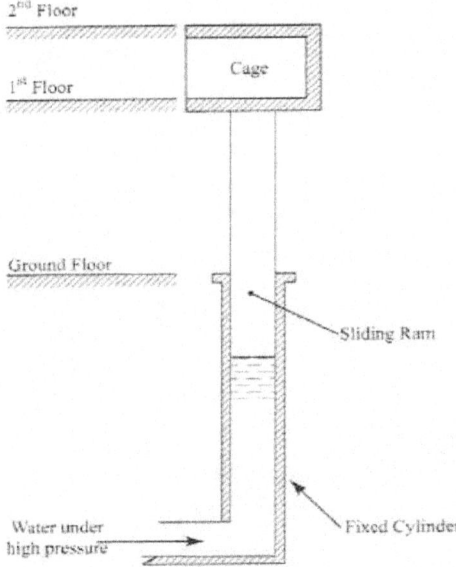

Fig (14-7) Direct acting hydraulic lift

A direct acting hydraulic lift consists of a cage (for placing the loads or standing passengers) firmly secured to the top of a vertical ram sliding in a fixed cylinder. There is a pipe connection for supplying water under a high pressure to the fixed cylinder as shown in Figure (14-7). The load to be lifted is placed in the cage of the lift. The water, under high pressure, is made to enter in the fixed cylinder which moves up the sliding ram and the cage fixed to it.

b) Suspended hydraulic lift:
A suspended hydraulic lift consists of a cage (for placing the loads or standing of passengers) which is suspended from a wire rope. The hydraulic lift obtains its motion from the jigger, in the same way as the hydraulic crane.

Near the foot of the cage hole, a jigger is provided. This jigger consists of a fixed cylinder having a pulley-block and containing a sliding ram. One end of the ram is in contact with the water and the other carries a pulley-block. A wire rope, one end of which is fixed, is taken round all the pulleys of the two blocks and finally over the guide pulleys. The cage is suspended from the other end of the rope. There is a pipe connection for supplying water under a high pressure to the fixed cylinder of the jigger as shown in Figure (14-8):

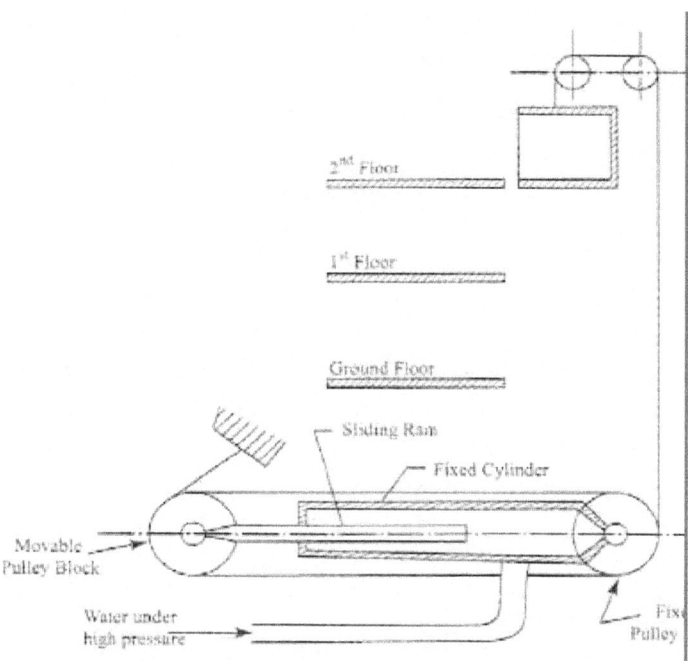

Fig (14-8) Suspended hydraulic lift

The load to be lifted is placed in the cage. The water under high pressure is admitted into the cylinder of the jigger. This water forces the sliding ram to move towards the left. The outward movement of the sliding ram makes the pulley-block move outwards. Due to increased distance between the two pulley-blocks, the wire rope is pulled and the cage is lifted up.

Check Your Knowledge

1) Explain, with help of a neat sketch, the working and the principle of hydraulic press.
2) Draw a neat sketch and explain the working of a hydraulic accumulator.
3) Define the capacity of a hydraulic accumulator.
4) By means of a neat sketch, explain the function and working of a hydraulic intensifier.
5) Explain with a sketch the working of a hydraulic ram and define efficiency of the ram.
6) Explain the working of the hydraulic crane with diagram.
7) Describe the working of hydraulic lift.

15. Introduction to Engineering Hydrology

15.1 Introduction

Hydrology denotes a field of science that covers many branches of science. In a much broader sense, hydrology refers to the study of water.

The term 'hydrology' is derived from two Greek words 'hydro' and 'logas' meaning 'water' and 'science', respectively. In simple terms, hydrology is a science related to water.

The US Federal Council for Science and Technology defines hydrology as:
'Hydrology is the science that treats waters on the earth, their occurrence, circulation, distribution, their chemical and physical properties and their environment including their relation to living things'

In short, what happens to the rain is the basis of the definition of the science of hydrology. It should not be confused with hydraulics, which deals with the mechanics of water.

As hydrology got a variety of practical applications, therefore, it should not be treated as a pure science.

Importance of hydrology: Water is the most valuable resource because human race or life will not survive in its absence. It not only supports the animal and plant kingdom for its daily subsistence but also serves as a valuable source of energy and a means of transportation. It also serves many other useful purposes.

However, this natural source, at times, assumes the form of a very destructive agent destroying valuable property, taking heavy toll of life and eroding and carrying thousands of tons of rich and fertile soil into the sea.

With rapid increase in population, the demands for this vital resource are becoming more and more acute. Also the destructive effects of floods are increasing and becoming more devastating. It is, therefore, necessary that an attempt be made to gain a better understanding of the occurrence and behaviour of water on earth.

15.2 Hydrologic Cycle

The hydrologic cycle begins with the evaporation of water from the surface of the ocean. As moist air is lifted, it cools and water vapour condenses to form clouds.

Moisture is transported around the globe until it returns to the surface as precipitation. When the water reaches the ground, one of the two processes may occur:
 (a) Some of the water may evaporate back into the atmosphere, or
(b) The water may penetrate the surface and become groundwater.

Groundwater either seeps its way to into the ocean, rivers, and streams, or is released back into the atmosphere through transpiration.

The balance of water that remains on the earth's surface is runoff, which empties into lakes, rivers and streams, where it is carried back to the oceans to start the cycle again.

A good example of the hydrologic cycle work is the snowfall lake effect. Figure (15-1) below explains the processes of hydrologic cycle that contribute to the production of lake effect shown:

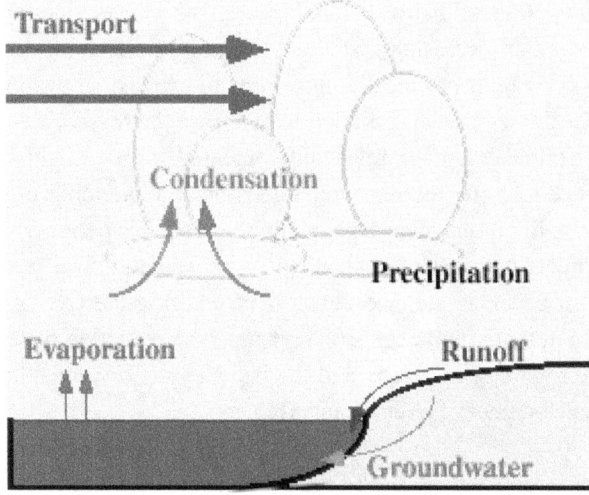

Fig (15-1) Lake effect hydrologic cycle

The cycle begins as cold winds (horizontal below arrow) blow across a large lake, a phenomena that occurs frequently in the late fall and winter months around great lakes.

Evaporation of warm surface water increases the amount of moisture in the colder drier air flowing immediately above the lake surface. With continued evaporation, water vapour in the cold air condenses to form ice-crystal clouds, which are transported toward the shore.

By the time these clouds reach the shoreline, they are filled with snow-flakes too large to remain suspended in the air and consequently, they fall along the shoreline as precipitation.

The intensity of lake effect snow-fall can be enhanced by additional lifting due to topographical features (hills) along the shoreline. Once the snow begins to melt, the water is either absorbed by the ground and becomes groundwater, or returns back to the lake as runoff.

15.3 Hydrological Cycle and Human Impact

Human activities that change the land cover of river basins and are aimed at regulating the water fluxes in nature can considerably change the hydrological cycle of the separate river basins, and even of large regions.

Due to human activities, the natural hydrological cycle of most river basins is becoming more and more transformed and regimented. The main stream flow regulation methods are construction of dams, levees, barrages, and dikes, which provide water accumulation, decreasing flood flow, and increasing low flow. The major effects of reservoir construction on the hydrological cycle (except runoff control) are an increase of evaporation and a rise of groundwater table. In dry regions evaporation losses from the reservoir water surface may be so large that they seriously compromise any potential gain. At the same time, in the conditions of moderate climate, the reservoir losses on evaporation are relatively small. On the other hand, the rise in groundwater level along the reservoir periphery and in surrounding areas changes the runoff generation mechanism on these areas. The gradual change of the river flow regime can occur as a consequence of decreasing the river's ability to transport sediments, especially in upper parts and in reservoirs. The reduction of sediment input at the dam site reduces the river channel slope and the bed sheer stress, resulting in dropping flow velocities and the development of river meandering.

In many dry regions, a considerable rise in the groundwater table can occur because of water filtration from reservoirs, leakage from water distributing systems, and faulty irrigation technology. Such a rise may cause water logging of plants and development of soil salinization.

To remove excess water from waterlogged soils, drainage is applied in many regions of the world. The primary effect of drainage is the lowering of the groundwater table and the extension of the layer with unsaturated soil. As a result, evapotranspiration may considerably drop. The improvement of hydraulic conditions due to drainage increase flow velocities. Acceleration of flow will also lead to a significant increase in flood peaks.

The effects of agricultural and forestry practices on the hydrological cycle are less apparent, and depend, to a significant extent, on the physiographic and climatic conditions. It is evident that ploughing, especially contour ploughing usually breaks up over land flow and increases infiltration

The main effects of deforestation on hydrological cycle of a river basin are the increases in transpiration and interception of precipitation, which in turn result in a decrease of the volume of total runoff. Deforestation reduces infiltration and improves the conditions for overland flow. As a consequence, flood runoff and peak discharges may significantly increase.

15.4 Precipitation

Precipitation is the water that falls from the atmosphere in either liquid or solid form. It results from the condensation of moisture in the atmosphere due to the cooling of a parcel of air. The most common cause of cooling is dynamic or adiabatic lifting of the air. Adiabatic lifting means that a given parcel of air is caused to rise with resultant cooling and possible condensation into very small cloud droplets. If these droplets coalesce and become sufficient size to overcome the air resistance, precipitation in some forms results.

15.4.1 Forms of precipitation: Precipitation occurs in various forms. Rain is precipitation that is in the liquid state when it reaches the earth. Snow is frozen water in a crystalline state, while hail is frozen water in a 'massive' state. Sleet is melted snow that is an intermixture of rain and snow. It should be recognized that precipitation that falls to earth in the frozen state cannot become part of the runoff process until melting occurs. Much of the precipitation that falls in mountainous areas and in north latitudes falls in the frozen form and is stored as a snowpack or ice until warmer temperatures prevail.

15.4.2 Types of precipitation (by origin): Precipitation can be classified by the origin of the lifting motion that causes the precipitation. Each type is characterized by different spatial and temporal rainfall regimes. The three major types of storms are classified as convective storms, orographic storms, and cyclonic storms. A fourth type of storm is often added, the hurricane or tropical cyclone:

(1) Convective storms:
Precipitation from convective storms results as warm moist air rises from lower elevations into cooler overlying air, as shown in Figure (15-2).

The characteristic form of convective precipitation is the summer thunderstorm. The surface of the earth is warmed considerably by mid-to late afternoon of a summer day, the surface imparting its heat to the adjacent air. The warmed air begins rising through the overlying air, and if proper moisture content conditions are met (condensation level), large quantities of moisture will be condensed from the rapidly rising, rapidly cooling air. The rapid condensation may often result in huge quantities of rain from a single thunderstorm spawned by convective action, and very large rainfall rates and depths are quite common beneath slowly moving thunderstorm.

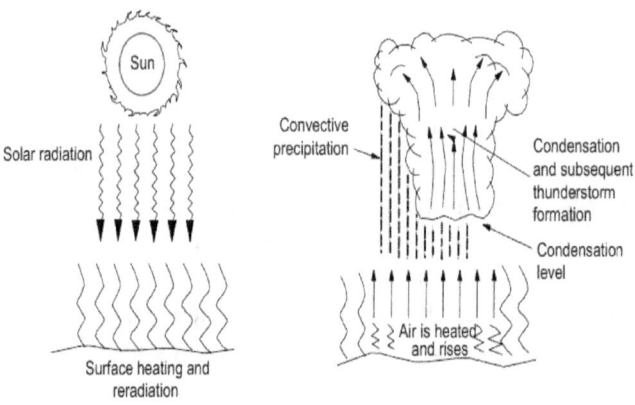

Fig (15-2) Convection storm

(2) Orographic storm:

Orographic precipitation results as air is forced to rise over a fixed-position geographic feature such as a range of mountains, as seen in Figure (15-3):

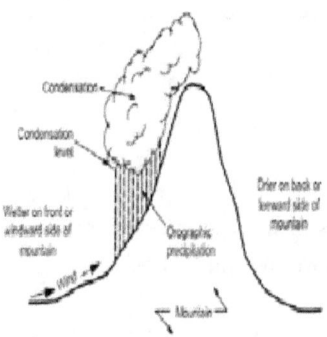

Fig (15-3) Orographic storm

(3) Cyclonic storm:

Cyclonic precipitation is caused by the rising or lifting of air as it converges on an area of low pressure. Air moves from areas of high pressure towards areas of low pressures. In the middle latitudes, cyclone storms generally move from west to east and have both cold and warm air associated with them. These mid-latitude cyclones are sometimes called extra-tropical cyclones of continental storms.

Continental storms occur at the boundaries of air of significantly different temperatures. A distribution in the boundary between the two air parcels can grow, appearing as a wave as it travels from west to east along the boundary. Generally, on a weather map, the cyclonic storm will appear as shown in Figure (15-4), with two boundaries or fronts developed:

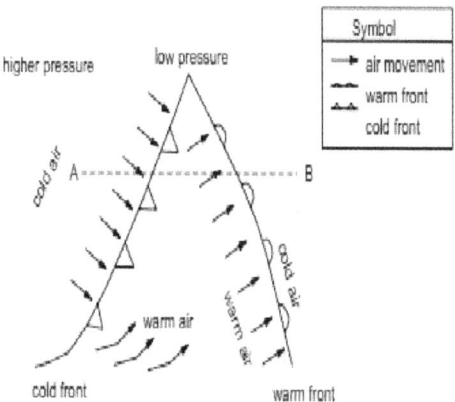

Fig (15-4) Storm as it appears on weather map in northern hemisphere

One of the boundaries has warm air being pushed into an area of cool air, while the other has cool air pushed into an area of warmer air. This type of movement is called a front; where warm air is the aggressor, it is a warm front, and where cold air is the aggressor, it is a cold front, as demonstrated in Figure (15-5):

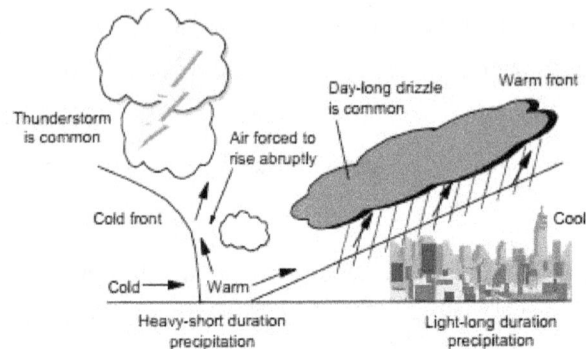

Fig (15-5) Cyclonic storm in mid-latitude; cross section from A to B of Figure (15-4)

The precipitation associated with a cold front is usually heavy and covers a relatively small area, whereas the precipitation associated with a warm front is more passive, smaller in quantity, but covers a much larger area.

(4) Hurricanes and typhoons:
Hurricanes, typhoons, or tropical cyclones develop over tropical oceans that have a surface-water temperature greater than 29°C (84°F). A hurricane has no trailing fronts, as the air is uniformly warm since the ocean surface from which it was spewed is uniformly warm.

Hurricanes can drop tremendous amounts of moisture on an area in a relatively short time. Rainfall and winds, often winds are sustained in excess of 120 km/h (75 miles/h) are common in well-developed hurricanes.

15.5 Evaporation, Transpiration and Evapotranspiration

Evaporation and transpiration are the primary abstractions of the hydrological cycle. These abstractions are small during a runoff event and can be neglected. The bulk of evaporation and transpiration takes place during the time between runoff events, which usually is long. Hence, these abstractions are the most important during this time interval. The combined effect of evaporation and transpiration is called 'evapotranspiration'.

Over a large land area in temperature zones, about two thirds of the annual precipitation is evapotranspired and the remaining one third runs off in streams and through the groundwater to the oceans.

In arid regions, evapotranspiration may be even more significant returning up to 90 percent or more of the animal precipitation to the atmosphere.

Evaporation also links hydrology to atmospheric science and, through transpiration, to agricultural sciences.

15.5.1 Evaporation: The process by which water is changed from the liquid or sold state into the gaseous state through the transfer of heat energy known as evaporation.

In the hydrological cycle evaporation is an important process, so much so that on a continental basis, approximately 70 to 75 per cent of the total annual precipitation is returned to the atmosphere by evaporation and transpiration. In hot climates, the loss of water by evaporation from rivers, canals and open-water storage equipment is a vital matter as evaporation takes a significant proportion of all the water supplies. It is significant in the sense that most of the water withdrawn for beneficial uses ultimately returns to streams and aquifers to become available for reuse, while the loss of water due to evaporation is entirely lost from the usable supply. Even in humid areas, evaporation loss is significant although the cumulative precipitation tends to mask it so that it is ordinarily not recognized except during rainless periods.

Strong reservoirs expose wide surfaces to evaporation and thus are a major source of water loss, even though they may lessen natural evaporation by confining floods in deep storages instead of spreading over a wide flood plains.

The factors controlling evaporation have known for a long time, but evaluating them is difficult because of their interdependent effects. However, in general, evaporation is affected by temperature, wind, atmospheric pressure, humidity, water quality, water depth, soil type and nature, and shape of the surface.

15.5.2 Transpiration: Transpiration is defined as a natural plant physiological process whereby water is taken from the soil moisture storage by roots and passes through the plant structure and is evaporated from cells in the leaf called stomata.

The amount of water held in storage by a plant is less than one per cent of that lost by it during the growing season. From the hydrological standpoint, therefore, plants are like pumps that remove water from the ground and raise it to the atmosphere.

It is difficult to make precise estimates of the water transpired because of the many variable responsible for the process. Available estimates should be used with due caution taking into consideration the conditions under which these estimates were obtained. Adequate relationships between climatic factors and transpiration are prerequisites if the data derived in one climatic region are supposed to have general utility.

Transpiration is affected by physiological and environmental factors. Stomata tend to open and close in response to environmental conditions such as light and dark, and heat and cold.

Environmental factors that affect transpiration are essentially the same as for evaporation, but can be considered a bit differently. For practical purposes, vapour pressure gradient, temperature, solar radiation, wind and available soil moisture are the most important factors

15.5.3 Evapotranspiration: The term evapotranspiration (ET) is defined as the water vapour produced from the watershed as a result of the growth of plants in the watershed.

Evapotranspiration and consumptive use include both the transpiration by vegetation and evaporation free surfaces, soil, snow, ice and vegetation. It is important to give the difference between evapotranspiration and consumptive use. Consumptive use differs from evapotranspiration only in that it includes the water used to make plant tissues. In computing evapotranspiration both transpiration and soil evaporation are included. The actual evapotranspiration can be determined by the analysis of the concurrent record of rainfall and runoff from a watershed.

There is an important difference between evapotranspiration and free surface evaporation. Transpiration is associated with plant growth and hence evapotranspiration occurs only when the plant is growing, resulting thereby in diurnal and seasonal variations. Transpiration thus superimposes these variations on the normal annual free water-surface evaporation.

15.6 Infiltration and Percolation

Infiltration is the downward movement of water that seeps into the soil or a porous rock.

Infiltration is controlled by soil texture, soil structure, vegetation and soil moisture status. High infiltration rates occur in dry soils, with infiltration slowing as the soil becomes wet. Coarse texture soils with large well-connected pore spaces tend to have higher infiltration rates than the fine textured soils. However, coarse textured soils fill more quickly than fine textured soils due to smaller amount of total pore space in a unit volume of soil. Runoff is generated quicker than one might have with a finer texture soil.

Vegetation also affects infiltration. Infiltration is higher for soils under forest vegetation than bare soils. Tree roots loosen and provide conduits through which water can enter the soil. Foliage and surface litter reduce the impact of falling rain keeping soil passages from becoming sealed.

Percolation is the movement of water through the soil or underlying porous rock. This water collects as groundwater.

Geologic formations in the earth crust serve as natural subterranean reservoirs for storing water. Others can also serve as conduits for the movement of water. Essentially, all groundwater is in motion; some of it, however, moves extremely slow. A geologic formation which transmits water from one location to another in sufficient quantity for economic development is called an aquifer. The movement of water is possible because of the voids or pores in the geologic formations. Some formations conduct water back to the ground surface. A spring is a place where the water table reaches the ground surface. Stream channels can be in contact with an unconfined aquifer that approaches the ground surface. Water may move from the ground into the stream, or vice versa, depending on the relative water level. Groundwater discharges into a stream forms the base flow of the stream during dry periods, especially during droughts. An influent stream supplies water to an aquifer while and effluent stream receives water from the aquifer.

15.7 Surface Runoff

Surface runoff is the water from rain, snowmelt, or other sources flowing over the land surface to ultimately reach streams. It is one of the major components in the water cycle. Surface runoff can be expressed in the water volume (or mass) per unit of area per unit of time.

If the amount of water falling on the ground is greater than the infiltration rate of the surface, runoff or overland flow will occur. Runoff specially refers to the water leaving an area of drainage and flowing across the land surface to points of lower elevation. It is not the water flowing beneath the surface of the ground.

Runoff involves the following events:
(a) Rainfall intensity exceeds the soil's infiltration rate and water begins accumulating at ground surface,
(b) Accumulating water causes a thin layer of water to form. This water layer begins to move downslope because of gravity,
(c) Flowing water accumulates into larger depressions on the ground surface,
(d) Depressions fill up and overflow forming small rills,
(e) Rills join to form larger streams and rivers, and
(f) Streams and rivers flow until they eventually empty into lakes or oceans.

On a global scale, runoff occurs because of the imbalance between evaporation and precipitation over the earth's land and ocean surface.

Check Your Knowledge

1) Define Hydrology? What is the importance of hydrology?

2) Explain the hydrologic cycle. Give examples of hydrologic cycle.

3) What is precipitation? Explain forms and types of precipitation.

4) Explain with aid of sketch: (a) convective storm, (b) Orographic storm, (c) Cyclonic storm, and (d) Hurricanes / typhoons.

5) Write short notes on: (a) Evaporation, (b) Transpiration, and (c) Evapotranspiration.

6) Define with examples both Infiltration and Percolation.

7) What events are involved with runoff?

16. Application of Engineering Hydrology

16.1 Introduction

Hydrology is the study of the movement, distribution, and quality of water on earth and other planets. A practitioner of hydrology is a hydrologist, working within the field of earth or environmental science, physical geography, geology or civil and environmental engineering.

The hydrologic problems fall into two broad categories:
(a) Estimating the size of the flood hydrograph, which may occur at infrequent intervals - or never. This governs the design of all flood control work and is important in all other problems.
(b) Estimating the extent to which river flow may drop during the summer, and the extent that this be regulated by storage.

In each case the study makes use of basic measurements of stream flow, of precipitation, temperature etc. Invariably all of these studies require an appreciation of the phases of the hydrologic cycle, although it is becoming only too obvious that the study of hydrology is a fulltime job in itself. In all cases the Civil Engineering is interested in the end result of the hydrologic cycle namely 'runoff'. To it is considered as losses whereas they may be thought of as gains by an irrigation engineer.

16.2 Predicting a Catchment Response to Rainfall

Relationships between catchment-scale hydrologic response and landscape attributes allow a prediction to be made of the hydrologic response of an ungagged catchment (a catchment which is not gauged for stream flow), based on climatological data and a description of the landscape attributes of that catchment. Development of such relationships is an example of a regionalisation methodology and is a useful goal for a number of reasons. For example, construction of a hydrological structure such as a bridge or dam may require prediction to be made of the hydrologic response catchment at an uncharged point.

For an examination of the ecological health of a river system, the frequency and duration of low flows may be of more interest, since these may be the limiting factors in supporting stream biota. If the catchment under consideration is not gauged for stream flow, these estimates must be based on some form of regionalisation, where the catchment is considered to behave similarly to another catchment (or catchments) with similar climatology and landscape attributes. These practical applications are the reason that many regionalisation studies deal with flood frequency analysis, or low flow analysis. Regionalisation is also of use in global and regional climate models, where the hydrologic response of an ungauged grid cell may be required for water and energy feedbacks in the land-atmosphere component of the climate model.

Several previous studies have related landscape attributes to the hydrologic response of a catchment as defined by a rainfall-runoff model. In general, regionalisation studies using a modelling methodology have met with only limited success. As a result, landscape attributes are commonly related directly to just one aspect of hydrologic response, such as food frequency.

16.3 The Unit Hydrograph Rainfall-run off Model

A unit hydrograph is the hypothetical unit response of a watershed (in terms of runoff volume and timing) to a unit input of rainfall. It can be defined as the 'direct runoff hydrograph' resulting from one unit (i.e. one cm or one inch) of effective rainfall occurring uniformly over that watershed at a uniform rate over a unit period of time.

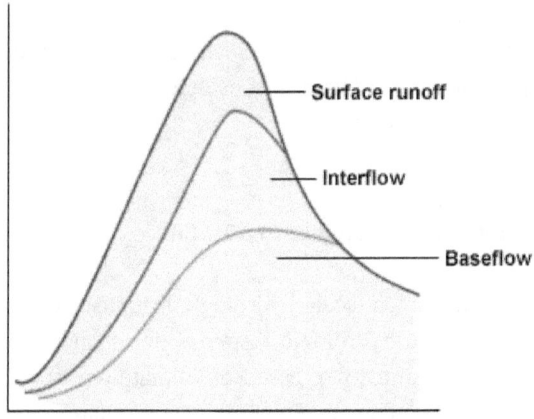

Fig (16-1) Surface flow runoff hydrograph

Some runoff models already account for the portion of quick-response runoff that occurs just below the surface, or interflow. Therefore, when using any runoff model, one should check to see whether the unit hydrograph needs to include interflow or any other portions of subsurface flow, so that these portions are not accounted for twice in the final output.

The unit hydrograph technique provides a practical and relatively easy-to-apply tool for quantifying the effect of a unit of rainfall on the corresponding runoff from a particular drainage basin. The unit hydrograph theory assumes that water shed's runoff response is linear and time-invariant, and that the effective rainfall occurs uniformly over the watershed. In the real world, none of these assumptions are strictly true. Nevertheless, application of unit hydrograph methods typically yields a reasonable application of the flood response of natural watersheds. The linear assumptions underlying unit hydrograph theory allows for the variation in storm intensity over time to be simulated by applying the principles of superposition and proportionality to separate storm components to determine the resulting cumulative hydrograph. This allows for a relatively straight forward calculation of the hydrograph response to any arbitrary rain event.

A unit hydrograph generated from a runoff model that has already separated both base flow and interflow in its computation will peak faster and higher than the more traditional unit hydrography, as shown in Figure (16-2):

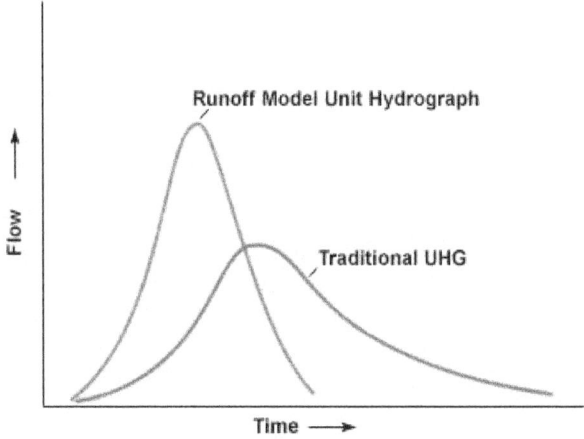

Fig (16-2) Standard vs runoff model unit hydrograph

This is because surface runoff will reach the stream faster than subsurface runoff, or interflow. Notice, however, that the area under the two curves is equal, because both hydrographs are accounting for the same, single unit of runoff.

16.4 Statical Analysis of River-flow Data

A major factor in the design of vehicular components is the anticipated severity in the usage of the component. In many situations, the reliability of the component reflects directly upon the safety of the vehicle, as well as the overall quality of the vehicle in the eye of the consumer, so designing for the anticipated service is of extreme importance. Quite often, in order to determine the loads that a component must be designed to withstand, extensive testing is conducted using a sampling of the population as a test base, and hundreds of sets of load data are measured. Typically, a component will be designed with the premise that it will have a life of least 'x' usage for some percentage of the population of users. Hence, one of the keys of design is determining what loading is associated with (x) % most extreme usage is.

Another factor in the design process is in verifying that the design life of the component is actually achievable for a large majority of the population. However, it is inconvenient and expensive to test them to failure when they are in service, because a typical component is designed to have a relatively long life. Instead, it would be convenient to conduct in-service testing on the part for a period of time much shorter than the expected life, and then use the loads that were found to occur in that period of time to predict the life of the component.

These two ideas for determining both the extended loads and the single most extreme loading scenario that a component will be subjected to are the basis for the river flow data statical analysis:

1) Information on the magnitude and frequency of occurrence of river flows is required for various evaluations, both economic and environmental, in relation to such things as water abstractions for domestic and industrial use, the formulation of licence conditions for effluent discharge during low flow, the design of bridges and drainage works, and flood prevention and alleviation during flood flows. In order to make these evaluations, flow records covering a considerable length of time must be available. Information is very often required at locations where no gauging has been carried out.

2) In the context of water resources management and pollution control, information is needed in different forms for various purposes:
- dry weather flow for abstraction,
- 95 per cent flow for effluent disposal,
- continuous pattern of flow for fisheries,
- instantaneous flow at a particular time, and
- groundwater flow to assess drought flow.

3) Provide details of the magnitude and frequency of occurrence of river flow based on records from gauging stations operated by various organisations for diverse purposes.

4) Data on stream flow are obtained by recording the water level at a gauge station on the river and by relating this level to the corresponding rate of flow. The precision of water level measurements and the sensitivity of the site to change in water level are factors which govern the accuracy and reliability of the data.

16.5 River-line and Surface Water Flood Risk Management

Floods are defined as a temporary covering by water land not normally covered by water. This term is generic term to include floods from rivers, mountain torrents, floods from the sea in coastal areas, and may exclude floods from sewage systems. Floods occur when the natural or man-made channels are unable to carry all the water, leading to rising water flows that flow over the banks and flood the surrounding dry land.

To ensure that flood risk from surface runoff, groundwater and small watercourses ditches, known as ordinary watercourses, to identified and managed as part of locally agreed work programmes.

16.5.1 River-line and coastal erosion: As all dams result in reduced sediment load downstream, a dammed river is said to be hungry for sediment. Because the rate of deposition of sediment is greatly reduced since there is less to deposit but the rate of erosion remains nearly constant, the water flow erodes the river shores and river beds, threatening shoreline ecosystem, deepening the river-bed, and narrowing the river over time. This leads to a compromised water table, reduced water levels, homogenization of the river flow and thus reduced ecosystem variability, reduced support for wildlife, and reduced amount of sediment reaching coastal plains and deltas. This prompts coastal erosion, as beaches are unable to replenish what waves erode without the sediment deposition of supporting river systems.

Channel erosion of rivers has its own set of consequences. The eroded channel could create a lower water table level in the affected area, impacting bottomland crops such as alfalfa or corn, and resulting in a smaller supply.

16.5.2 Surface water flood risk management: Flooding is defined as the inundation of land by water. The magnitude of the flood risk is dependent on the probability of a flood occurring, the value and type of assets or resources exposed to the risk, and the vulnerability of those assets or resources to damage.

The effects of flooding include the movement of debris, the building-up of debris against structures, silt and / or mud deposition, erosion, and water damage to buildings and vehicles. Consequently contamination and health effects may arise from overloaded sewerage systems or transportation of hazardous substances.

Flooding can be caused by a range of factors and circumstances including:
- High, or particularly intense, periods of rainfall,

- Snow melt (which may also coincide with high rainfall), and
- Back water ways or drainage systems (including natural damming after landslips or earthquake, or vegetation blocking drains, creeks or streams).

Human activity can also contribute to, or exacerbate, flood hazard by, for example:
- Obstructing natural overland flow paths (such as by placing buildings, raised roadways, embankments and other similar obstacles in the flow path or flood channel), and
- Increasing the flow of water into natural or manmade drainage systems (removing vegetation, increasing areas of impermeable surfaces, or increasing the number of storm water outlets, and thereby the amount of storm water, that enters particular drainage system).

There are many ways to mitigate the flood risk, but generally they fall into two groups:
- Structural work: designed to contain floods and to limit erosion and deposition by controlling river behaviour, and
- Non-structural methods: including land-use planning, emergency management planning, and flood-proofing of buildings. These methods are designed to either remove people and assets from risk or to manage exposure to flood effects.

16.6 Sewer Design Using Rational and Modified Rational Method

A sewer system is a network of pipes used to convey storm runoff and/or wastewater in an area. The design of sewer system involves the determination of diameters, slopes, and crown or invert elevations for each pipe in the system.

16.6.1 Rational method: The rational method uses an empirical linear equation to compute the peak runoff rate from a selected period of uniform rainfall intensity.

Originally developed more than 100 years ago, it continues to be useful in estimating runoff from simple, relatively small drainage areas such as parking lot.

Use of the rational method should be limited to drainage areas such as parking lots. They are limited to drainage areas less than 20 acres with generally uniform surface cover and topography.

It is important to note that the rational method can be used only to compute peak runoff rates. Since it is not based on total storm duration, but rather a period of rain that produces the peak runoff rate, the method cannot compute runoff volumes unless the user assumes a total storm duration.

16.6.2 Modified rational method: The modified rational method is a somewhat recent adaptation of the rational method that can be used to not only compute peak runoff rates, but also to estimate runoff volumes and hydrographs. This method uses the same input data and coefficients as the rational method along with further assumption that, for the selected storm frequency, the duration of peak-producing rainfall is also the entire storm duration.

Since, theoretically, there are an infinite number of rainfall intensities and associated durations with the same frequency or probability, the modified rational method requires that several of these events be analysed in the method to determine the most severe. Use of modified rational method should also be limited to drainage areas less than 20 acres with generally uniform surface cover and topography.

16.7 Water Supply Reservoir

Man-made reservoirs, or dams, are purpose built principally to provide water supply to homes, industry or agriculture or, in some cases, for electric power generation.

16.7.1 Water storage: There are three basic locations of water storage in the planetary water cycle. Water is stored in the atmosphere, water is stored on the surface of the earth, and water stored in the ground.

Water stored in the atmosphere can be moved relatively quickly from one part of the planet to another part of the planet. The type of storage that occurs on the land surface and under the ground largely depend on the geologic features related to the types of soil and the types of rocks present at the storage locations. Storage occurs as surface storage in oceans, lakes, reservoirs, and glaciers; underground storage occurs in the soil, in aquifers, and in the crevices of rock formations.

16.7.2 Reservoir design: Water utilities own and operate a number of portable water reservoirs throughout their water supply network. Reservoirs help providing onsite storage to buffers for peak demands and emergency supply situations. However increased residence times in reservoirs result in residual chlorine decay that augments the risk of recontamination in the network and can significantly affect water quality in distribution systems.

Reservoirs design and operation are critical to minimize water age. It is essential to size reservoirs such that residence time is as low as possible whilst presenting sufficient volume to provide security of supply. Volume of water flowing in and out of the reservoirs must be maximised and reservoirs must be regularly filled and emptied through network operation to guarantee renewal of the water stored.

The behaviour of water in reservoirs is another decisive parameter that can impact on water quality. It is important to understand the movement of water within water tanks in order to ensure that portable water is not held up in 'dead zones'. Ideal flow pattern through a reservoir is plug flow. In plug flow, all fluid elements entering the reservoir have a constant velocity and present the same hydraulic residence time. In these conditions water passing through the reservoir is not retained in 'dead areas' and water age is minimised.

Check Your Knowledge
1) Define hydrology? Mention two categories for hydrologic problems.
2) What are the factors that relate to the prediction of catchment response to rainfall?
3) Describe the unit hydrograph rainfall run off model.
4) Mention the basis for a statical analysis of river flow data.
5) What is river-line and coastal erosion? Explain surface water flood risk management.
6) Define sewer design using rational and modified rational methods.
7) What are the three basic locations of water storage?
8) What is the importance behind reservoir design and its operation?

17. Sustainable Drainage System (SUDS)

17.1 Introduction

Whilst Sustainable Drainage Systems (SUDS) have been around for some time, their use is becoming increasingly important due to recognition of the need to accommodate water runoff from hard surfaces in a more effective and cost efficient manner that also brings benefits to the environment.

The SUDS concept is to mimic, as closely as possible, natural drainage of a site in order to minimise the impact that urban development has on flooding and pollution of rivers, streams and other water bodies.

The use of a variety of techniques within the management train allows the SUDS concept to be applied to all sites. The techniques utilising vegetative features to treat pollution and slow down or reduce flows can enhance landscape and provide wildlife habitat.

17.2 What are Sustainable Drainage Systems?

Sustainable drainage systems (SUDS) are increasingly being used to mitigate the flows and pollution from runoff. The philosophy of SUDS is to replicate as closely as possible the natural drainage from a site before development and to treat runoff to remove pollutants, so reducing the impact on receiving water courses. This requires a reduction in the rate and volume of runoff from developments, combined with treatment to remove pollutants as close to the source as possible. They can also provide other environmental benefits such as wildlife habitat, improve aesthetics or community resource.

SUDS permit a very flexible approach to be taken to drainage, and the techniques available range from soak ways to large-scale detention basins. The individual techniques are used in series in a management train designed to meet the site specific constraints. The techniques are not new, and may have been successfully used worldwide for more than 20 years. A wealth of knowledge about their performance has been developed. Over the past decade, a comprehensive SUDS research and monitoring programme was conducted, which is beginning to yield ample performance data on systems with global climate.

Some common misconceptions about SUDS and what they comprise include:
- SUDS is the use of soak ways,
- SUDS cannot be used on clay soils,
- SUDS is the use of ponds and wet lands,
- SUDS is storing rainwater on site and allowing it to flow at a restricted rate, and
- SUDS do not include pipes.

None of these statements is entirely correct. The SUDS approach to drainage involves controlling the runoff from development sites so that it mimics greenfield runoff and maintains the natural drainage patterns, as far as possible. SUDS should also enhance the local environment.

To achieve this, a treatment or management train is required that comprises one or more techniques. They may or may not include soak ways, ponds and wetlands or pipes. The management train may also include techniques such as good site management to prevent pollution. Several SUDS techniques will be needed to reduce the volume of runoff and treat pollution.

A drainage set-up that does not provide a management train to meet all three criteria of quality, quantity and amenity may not be a sustainable drainage system in the strictest sense, although on some sites specific factors it may be that one criterion is more prominent than the others. A SUDS approach to drainage can and should be applied to all sites, although site constraints may limit the potential for a truly sustainable solution.

Sustainable drainage systems may also incorporate storage for water reuse. Here the permanent storage volume will generally be additional to any storage volume required to control runoff rates, unless a continuous rate of use can be generated.

17.3 Design of SUDS

When designing SUDS, one should consider the three following elements:
- Quantity-Amount of water/design storm to be catered for,
- Quality-Improvement in water quality/removal of pollutants, and
- Biodiversity/Amenity-Improvement in habitat and environments contribution to new habitat/environment.

In order to design a successful SUD scheme for a site, it is best to provide for all above elements. The design of SUDS should provide for coherent management of the runoff water, including the different stages of treatment, which can comprise various techniques to achieve the required outcome.

A successful SUD scheme does not necessarily require infiltration into the subsoil. SUDS can include infiltration, attenuation or a hybrid of both. Where infiltration is used it is important to know the groundwater levels for the site, which should be monitored over a 12 months period, prior to development / construction.

17.3.1 Design procedure:

Step 1: For greenfield sites SUDS are designed to mimic greenfield runoff rates (i.e. such that the site acts as though it were a natural green site). For brownfield sites SUDS must be designed to mimic the existing runoff rates, including allowing for climate change, as a minimum but ideally to mimic green field rates.

Step 2: Existing runoff rates or green-field runoff rates must be calculated based on the site area, area of impermeable surfaces and geology. Calculation of SUDS can be complex as different storm durations and scale (up to a 1% probability rain storm) must be considered as well as the permeability of the ground conditions, the height of the water-table, and the impact of climate change.

Step 3: Where possible SUDS should allow water to filter into the ground (infiltration), mimicking natural processes.
However, if the site has been used for industry it may contain pollutants, therefore making infiltration inappropriate. Equally if the ground is predominantly clay or there is a high water table, the runoff rate may be so low that permeable solutions may be ineffective.

Step 4: Hard surfaces such as roofs and roads result in increased runoff rates over green surfaces. Green roofs and permeable pavements will help to reduce the runoff rates closer to green-field rates. Systems that slow the rainfall before it gets to the ground, such as green or brown roofs and green walls, are also universally beneficial and will continue to reduce runoff rates even in flood-prone areas.

Step 5: If it is not possible to use infiltration devices, and/or permeable surfaces are insufficient to reduce runoff rates sufficiently, then flood storage devices may need to be considered; these can include dry ponds swales, basins but also underground storage (typically pumped).
Communal rainwater harvesting tanks may be considered, however they are often far smaller than required for large rainstorms.

Step 6: Ground level SUDS such as swales or rain gardens can achieve water quality and environmental improvements, through habitat creation or amenity provision. When interspersed between parking areas and buildings they can also create visual interest. However, maintenance and protection of use must be considered.

17.4 Potential Problems with SUDS

To understand SUDS we must first consider the natural hydrological behaviour of a green-field site. When rain falls on such a site, it normally soaks into the soil. Only where there is particularly heavy rainfall will some of it run off slowly over the ground surface to the nearest ditch or watercourse. Most pollutants are filtered through soils or broken down by bacteria.

When these green-field sites are built on, much of the area becomes impermeable. With no soakage available, runoff is piped to the nearest watercourse or storm drain. Thus both the volume and rate of runoff can dramatically increase.

Inbuilt up urban areas this runoff ends up in urban streams or existing pipes that were never designed for these loads. This may lead to flooding or increased overflow from combined sewer, neither of which is acceptable. A combined sewer is where foul and surface water flows are transported in the same pipe. Excess runoff also causes problems with increased surcharging of pipes. This causes pipe damage and maintenance problems and may lead to foul sewage escaping from the pipe into the groundwater. Excess flows in combined systems can also lead to increased costs for wastewater treatment. The potential to naturally remove pollutants is lost.

SUDS are defined as a sequence of management practices and control structures designed to drain surface water in a more sustainable fashion than some conventional techniques: Using SUDS techniques, water is either infiltrated or conveyed more slowly to watercourses via ponds, swales or other installations.

The use of SUDS closely mimics natural catchment behaviour, results in attenuation of storm-water runoff, and improved environmental performance. SUDS systems vary from infiltration trenches/soak ways filter drains and permeable pavements to swales, detention basin and storm water wetland.

Check Your Knowledge

1) Explain the SUDS concept and techniques.
2) What are sustainable drainage systems?
3) What are the three elements to be considered when designing SUDS?
4) Mention the six steps for SUDS.
5) Indicate potential problems with SUDS.

Selected References

1. Bansal, R.K; (2005): "Fluid Mechanics and Hydraulic Machines," Laxmi Publication (P) LTD, New York

2. Brater E., King H., Lindell J., and Wei C; (1996): "Handbook of Hydraulics," Seventh Edition - McGraw Hill Company -USA- ISBN 0-07-007247-7.

3. Crowther J.M., and Dandy G.C; (2010): "Simulation of water flow and traces experiments for a clear water storage tank," Proceedings of First European Congress of the IAHR, Edinburgh, UK 4-6 May 2010, pp1-6.

4. Das, M.M., Saika, M.D; (2009): "Hydrology," Phi Learning Private Limited, New Delhi.

5. Dingman, S.L; (2003): "Physical Hydrology - 2nd edition-," AMAZON Publication.

6. Fetter, C.W; (2014): "Applied Hydrogeology," Pearson Education Limited - UK.

7. Gregory M., Jiohua, F; (1998): "Reservoir Sedimentation Handbook," McGraw-Hill Publishers.

8. Hamill L; (2011): "Understanding Hydraulics," Palgrave Macmillan - UK - ISBN: 0230242758.

9. Hunt, T., Vaughan, N; (1996): "Hydraulic Handbook," Elsevier Science Ltd, published by: Elsevier Advanced Technology, UK.

10. King, H.W., Wissler, C.O; (1933): "Hydraulics," J. Wiley - UK.

11. Marriot, M; (2009): "Civil Engineering Hydraulics," John Wiley & Son - UK.

12. Northeast National Technical Center; (June 1986): "Hydrology Technical," USDA National Resources Conservation Service.

13. Parr, A; (1998): "Hydraulic and Pneumatics," Elsevier Ltd, - Butterworth Heinemann Publication -UK.

14. Pippenger, J.J., Hicks, T.G; (1979): "Industrial Hydraulics," McGraw Hill - USA.

15. Raghunath, H.M; (2006): "Hydrology, Principles, Analysis, Design," New Age International Publisher - New Delhi.

16. Roberson, J.A., Cassidy, J.J., Chaudhry, M.H; (2005): "Hydraulic Engineering," - Second Edition - John Wiley & Sons. Inc. USA ISBN: 0-471-12466-4.

17. Russel, G.E; (1918): "Text Book on Hydraulics," Henery Holt and Company - New York.

18. Wanielista, M.P, Yousef, Y.A; (1993): "Storm Water Management," John Wiley & Sons, Inc. New York.

19. Whipple, W, Grigg, N.S, Grizzard, T, Randall C. W, Shubinski R.P, Tucker, L.S; (1983): "Storm Water Management in Urbanizing Areas," Prentice - Hall, Inc. Englewood Cliffs, New Jersey.

20. Zaher, M.A; (2011): "Principles of Cavitation and Bleeding System Application," AMAZON publication – ISBN: 1463768044

INDEX   Page

D

www.ingramcontent.com/pod-product-compliance
Lightning Source LLC
Chambersburg PA
CBHW080237180526
45167CB00006B/2313